Abstract Lie Algebras

The MIT Press Cambridge, Massachusetts, and London, England

Abstract Lie Algebras

David J. Winter

This book was designed by The MIT Press Design Department.
It was set in Monotype Times New Roman.

Printed and bound by The Colonial Press Inc.,
in the United States of America.

ISBN 0 262 23 051 8 (hardcover)

Library of Congress catalog card number: 70–165073

Contents

Preface

This book grew out of a one-semester course on Lie algebras given at the University of Michigan in 1968–1969. Aside from basic algebra and linear algebra, the material is self-contained. The first three chapters may be regarded as a solid introduction to the theory of Lie algebras. They may also be of interest to those who are already familiar with Lie algebras, since the development of the theory and the proofs quite often are not along standard lines. The fourth chapter consists mainly of material of fairly recent origin, including some unpublished material, on the general structure of Lie algebras of arbitrary characteristic.

Chapter 1 contains the prerequisites on modules. In Chapter 2, basic material is developed which is most naturally formulated for nonassociative algebras. This material is fairly standard, except that some results on characteristic ideals are discussed in 2.4.17, 2.6.7, 2.6.8, and a non-associative algebra version of an important theorem of Hans Zassenhaus is given in 2.7.5. Chapter 3 consists of a brief account of the highlights of the theory of Lie algebras of characteristic 0. In Chapter 3, I have attempted to use only the simplest and most direct kinds of arguments, deferring more general methods until Chapter 4. In 3.6, the proofs of Levi's theorem and the complete reducibility of representations of semi-simple Lie algebras are proved using a "Fitting's lemma" for semisimple Lie algebras. In 3.7, the theory of root systems is developed and applied to the classification of split semisimple Lie algebras. The proof of the isomorphism of split semisimple Lie algebras with isomorphic root systems is proved using a "diagonal subalgebra" argument that is similar to a "diagonal submodule" argument used in proving the isomorphism of irreducible modules having the same highest weight. In 3.8, auto-morphisms of Lie algebras are studied, and the conjugacy theorems on Cartan subalgebras and maximal solvable subalgebras are proved by algebraic (nongeometric) methods based on techniques developed by George Daniel Mostow. Chapter 4 consists of a development of material on the structure of Lie algebras of arbitrary characteristic. This structure is studied in terms of gradings of the Lie algebra and in terms of special subalgebras. In 4.4, Engel subalgebras and Fitting subalgebras are intro-duced and studied. These are then used in an extensive study of Cartan subalgebras. In 4.5 and 4.6, the distribution of Cartan subalgebras and maximal tori in a Lie p-algebra is discussed.

In developing parts of 3.7.3 on abstract root systems, I was heavily influenced by lectures given by Jacques Tits at Yale University in 1966–1967. Several other ideas used in that section arose out of conversations with Forrest Richen and James Humphreys. Ideas due to George Seligman thread their way through the book. Much of the material on automorphisms has its roots in papers of Armand Borel and George Daniel Mostow. Theorem 4.3.7 of Nathan Jacobson, together with a short, but incisive, related paper [12] of his, form the starting point for much of Chapter 4.

David J. Winter
Ann Arbor, Michigan
April 1971

Abstract Lie Algebras

1

Modules

1.1 Introduction

In this chapter, we introduce the language of modules in a form designed for the material developed later in the book. Throughout, $\mathfrak{S}$ is a set and k a field.

We begin with some basic definitions and properties of modules.

1.1.1 Definition

An $\mathfrak{S}$-*module* over k is a vector space $\mathfrak{M}$ over k together with a mapping $\mathfrak{M} \times \mathfrak{S} \to \mathfrak{M}$, denoted $(m, s) \mapsto ms$, such that $(\alpha m + \beta n)s = \alpha(ms) + \beta(ns)$ for $\alpha, \beta \in k$, $m, n \in \mathfrak{M}$, and $s \in \mathfrak{S}$.

1.1.2 Definition

Let $\mathfrak{M}$ be an $\mathfrak{S}$-module over k, $T \in \mathfrak{S}$. Then T_M is the linear transformation of $\mathfrak{M}$ defined by $mT_M = mT$ for $m \in \mathfrak{M}$.

1.1.3 Definition

The *direct sum* of $\mathfrak{S}$-modules $\mathfrak{M}$, $\mathfrak{N}$ over k is the $\mathfrak{S}$-module with underlying vector space the direct sum $\mathfrak{M} \oplus \mathfrak{N}$ of the vector spaces $\mathfrak{M}$, $\mathfrak{N}$ together with the mapping $(\mathfrak{M} \oplus \mathfrak{N}) \times \mathfrak{S} \to \mathfrak{M} \oplus \mathfrak{N}$ defined by

$((m + n), s) \mapsto ms + ns$ for $m \in \mathfrak{M}$, $n \in \mathfrak{N}$, and $s \in \mathfrak{S}$.

For $\mathfrak{M}$ an $\mathfrak{S}$-module over k and $\mathfrak{N} \subset \mathfrak{M}$, we let $\mathfrak{N}\mathfrak{S}$ be the subspace of $\mathfrak{M}$ generated by the set $\{ns \mid n \in \mathfrak{N}, s \in \mathfrak{S}\}$.

1.1.4 Definition

An $\mathfrak{S}$-*submodule* of an $\mathfrak{S}$-module $\mathfrak{M}$ is a subspace $\mathfrak{N}$ of $\mathfrak{M}$ such that $\mathfrak{N}\mathfrak{S} \subset \mathfrak{N}$. Such an $\mathfrak{N}$ is *proper* if $\mathfrak{N} \neq \mathfrak{M}$ and $\mathfrak{N} \neq \{0\}$.

1.1.5 Definition

If $\mathfrak{N}$ is an $\mathfrak{S}$-submodule of an $\mathfrak{S}$-module $\mathfrak{M}$, the *quotient* $\mathfrak{S}$-*module* $\mathfrak{M}/\mathfrak{N}$ of $\mathfrak{M}$ by $\mathfrak{N}$ is the vector space quotient $\mathfrak{M}/\mathfrak{N}$ together with the mapping $\mathfrak{M}/\mathfrak{N} \times \mathfrak{S} \to \mathfrak{M}/\mathfrak{N}$ defined by $(m + \mathfrak{N}, s) \mapsto ms + \mathfrak{N}$ for $m \in \mathfrak{M}$, $s \in \mathfrak{S}$.

1.1.6 Definition

An $\mathfrak{S}$-*homomorphism* from an $\mathfrak{S}$-module $\mathfrak{M}$ over k to an $\mathfrak{S}$-module $\mathfrak{N}$ over k is a linear transformation $f: \mathfrak{M} \to \mathfrak{N}$ over k such that $f(ms) = f(m)s$ for $m \in \mathfrak{M}$, $s \in \mathfrak{S}$. The set of such f is denoted $\mathrm{Hom}_k^{\mathfrak{S}}(\mathfrak{M}, \mathfrak{N})$. If such an f is bijective, it is an $\mathfrak{S}$-*isomorphism*.

Note that the canonical vector space homomorphism $\mathfrak{M} \to \mathfrak{M}/\mathfrak{N}$ is an $\mathfrak{S}$-homomorphism for $\mathfrak{M}$ an $\mathfrak{S}$-module and $\mathfrak{N}$ an $\mathfrak{S}$-submodule of $\mathfrak{M}$. One has the following fundamental homomorphism theorems, which are straightforward generalizations of the usual theorems.

1.1.7 Theorem

Let $f: \mathfrak{M} \to \mathfrak{N}$ be an $\mathfrak{S}$-homomorphism of $\mathfrak{S}$-modules $\mathfrak{M}, \mathfrak{N}$. Then the subspaces Kernel f and Image f are $\mathfrak{S}$-submodules of $\mathfrak{M}$ and $\mathfrak{N}$ respectively, and there is a canonical $\mathfrak{S}$-isomorphism from $\mathfrak{M}/\text{Kernel } f$ to Image f given by $m + \text{Kernel } f \mapsto f(m)$ for $m \in \mathfrak{M}$.

1.1.8 Theorem

Let $\mathfrak{P}, \mathfrak{Q}$ be $\mathfrak{S}$-submodules of an $\mathfrak{S}$-module $\mathfrak{M}$. Then there is a canonical $\mathfrak{S}$-isomorphism from $\mathfrak{Q}/\mathfrak{P} \cap \mathfrak{Q}$ to $\mathfrak{P} + \mathfrak{Q}/\mathfrak{P}$ given by $q + (\mathfrak{P} \cap \mathfrak{Q}) \mapsto q + \mathfrak{P}$ for $q \in \mathfrak{Q}$.

1.2 Complete Reducibility

We now let $\mathfrak{M}$ be any $\mathfrak{S}$-module over k.

1.2.1 Definition

An $\mathfrak{S}$-*complement* of an $\mathfrak{S}$-submodule $\mathfrak{N}$ of $\mathfrak{M}$ is an $\mathfrak{S}$-submodule $\mathfrak{N}'$ of $\mathfrak{M}$ such that $\mathfrak{M} = \mathfrak{N} \oplus \mathfrak{N}'$.

1.2.2 Definition

$\mathfrak{M}$ is $\mathfrak{S}$-*completely reducible* if $\mathfrak{M}\mathfrak{S} = \mathfrak{M}$ and every $\mathfrak{S}$-submodule of $\mathfrak{M}$ has an $\mathfrak{S}$-complement.

1.2.3 Proposition

Let $\mathfrak{M}$ be $\mathfrak{S}$-completely reducible, $\mathfrak{N}$ an $\mathfrak{S}$-submodule of $\mathfrak{M}$. Then $\mathfrak{N}$ and $\mathfrak{M}/\mathfrak{N}$ are $\mathfrak{S}$-completely reducible.

PROOF. Let $\mathfrak{P}$ be an $\mathfrak{S}$-submodule of $\mathfrak{N}$, and let $\mathfrak{P}'$ be an $\mathfrak{S}$-complement of $\mathfrak{P}$ in $\mathfrak{M}$. Then $\mathfrak{M} = \mathfrak{P} \oplus \mathfrak{P}'$ and it follows that $\mathfrak{N} = \mathfrak{P} \oplus (\mathfrak{P}' \cap \mathfrak{N})$. Furthermore, $\mathfrak{N}\mathfrak{S} = \mathfrak{N}$ since $\mathfrak{N}\mathfrak{S} + \mathfrak{N}'\mathfrak{S} = \mathfrak{M}\mathfrak{S} = \mathfrak{M} = \mathfrak{N} \oplus \mathfrak{N}'$. Thus $\mathfrak{N}$ is $\mathfrak{S}$-completely reducible.

Next, let $\overline{\mathfrak{P}}$ be an $\mathfrak{S}$-submodule of $\overline{\mathfrak{M}} = \mathfrak{M}/\mathfrak{N}$. Let $\mathfrak{P}'$ be an $\mathfrak{S}$-complement of the inverse image $\mathfrak{P}$ of $\overline{\mathfrak{P}}$ under the canonical $\mathfrak{S}$-homomorphism $\mathfrak{M} \to \overline{\mathfrak{M}}$. Then $\mathfrak{M} = \mathfrak{P} \oplus \mathfrak{P}'$, so $\overline{\mathfrak{M}} = \overline{\mathfrak{P}} \oplus \overline{\mathfrak{P}'}$ where $\overline{\mathfrak{P}'}$ is the image of $\mathfrak{P}'$ under the canonical $\mathfrak{S}$-homomorphism $\mathfrak{M} \to \overline{\mathfrak{M}}$. Obviously, $\mathfrak{M}\mathfrak{S} = \mathfrak{M}$ implies that $\overline{\mathfrak{M}}\mathfrak{S} = \overline{\mathfrak{M}}$. Thus, $\overline{\mathfrak{M}}$ is $\mathfrak{S}$-completely reducible.

It is very much not the case that if $\mathfrak{N}$ and $\mathfrak{M}/\mathfrak{N}$ are $\mathfrak{S}$-completely reducible, then $\mathfrak{M}$ is $\mathfrak{S}$-completely reducible. For instance, let $\mathfrak{V}$ be a finite-dimensional vector space over k of dimension at least 2, $\mathfrak{W}$ a subspace of $\mathfrak{V}$ of codimension 1, and let $\mathfrak{S} = \{T \in \operatorname{Hom}_k \mathfrak{V} \mid \mathfrak{W}T \subset \mathfrak{W}\}$. Then $\mathfrak{V}/\mathfrak{W}$ and $\mathfrak{W}$ are $\mathfrak{S}$-completely reducible, but $\mathfrak{V}$ is not. More generally, an $\mathfrak{S}$-module $\mathfrak{V}$ of the following kind is not $\mathfrak{S}$-completely reducible. We let $\mathfrak{V}$ be a vector space over k together with a chain $\mathfrak{V} = \mathfrak{V}_1 \supset \mathfrak{V}_2 \supset \ldots \supset \mathfrak{V}_n = \{0\}$ of $n \geq 3$ distinct subspaces, define $\mathfrak{S} = \{T \in \operatorname{Hom}_k \mathfrak{V} \mid \mathfrak{V}_i T \subset \mathfrak{V}_i \text{ for } 1 \leq i \leq n\}$ and let $\mathfrak{V} \times \mathfrak{S} \to \mathfrak{V}$ be defined by $(v, T) \mapsto vT$ for $v \in \mathfrak{V}$, $T \in \mathfrak{S}$.

1.2.4 Definition

$\mathfrak{M}$ is $\mathfrak{S}$-*irreducible* if $\mathfrak{M}\mathfrak{S} = \mathfrak{M}$ and $\mathfrak{M}$ has no proper $\mathfrak{S}$-submodules.

1.2.5 Lemma

Every nonzero $\mathfrak{S}$-completely reducible module $\mathfrak{M}$ has a nonzero $\mathfrak{S}$-irreducible $\mathfrak{S}$-submodule.

PROOF. Let x be a nonzero element of $\mathfrak{M}$. By Zorn's lemma, there is a maximal $\mathfrak{S}$-submodule $\mathfrak{N}$ of $\mathfrak{M}$ such that $x \notin \mathfrak{N}$. Now $\mathfrak{M} = \mathfrak{N} \oplus \mathfrak{N}'$ for some nonzero $\mathfrak{S}$-submodule $\mathfrak{N}'$ of $\mathfrak{M}$. But $\mathfrak{N}'$ is $\mathfrak{S}$-irreducible, for otherwise $\mathfrak{N}' = \mathfrak{N}'' \oplus \mathfrak{N}'''$ for suitable proper $\mathfrak{S}$-submodules $\mathfrak{N}''$, $\mathfrak{N}'''$. But then $x \in (\mathfrak{N} + \mathfrak{N}'') \cap (\mathfrak{N} + \mathfrak{N}''')$ by the maximality of $\mathfrak{N}$, so that $x \in \mathfrak{N}$, a contradiction.

1.2.6 Theorem

The following conditions are equivalent.

1. $\mathfrak{M}$ is $\mathfrak{S}$-completely reducible;
2. $\mathfrak{M} = \sum \mathfrak{M}_a$ for some collection $\{\mathfrak{M}_a \mid a \in A\}$ of $\mathfrak{S}$-irreducible S-submodules of $\mathfrak{M}$;
3. $\mathfrak{M} = \sum \oplus \mathfrak{M}_a$ for some collection $\{\mathfrak{M}_a | a \in A\}$ of $\mathfrak{S}$-irreducible $\mathfrak{S}$-submodules of $\mathfrak{M}$.

PROOF. We show that $1 \Rightarrow 2 \Rightarrow 3 \Rightarrow 1$. Thus, assume that $\mathfrak{M}$ is $\mathfrak{S}$-completely reducible and let $\mathfrak{N} = \sum \mathfrak{M}_a$ where $\{\mathfrak{M}_a | a \in A\}$ is the collection of all $\mathfrak{S}$-irreducible $\mathfrak{S}$-submodules of $\mathfrak{M}$. Then $\mathfrak{M} = \mathfrak{N} \oplus \mathfrak{N}'$ for some $\mathfrak{S}$-submodule $\mathfrak{N}'$ of $\mathfrak{M}$. Clearly $\mathfrak{N}'$ has no nonzero $\mathfrak{S}$-irreducible $\mathfrak{S}$-submodule. Thus, $\mathfrak{N}' = \{0\}$ by 1.2.3 and 1.2.5. Thus, $\mathfrak{M} = \sum \mathfrak{M}_a$ and $1 \Rightarrow 2$. Next, assume that $\mathfrak{M} = \sum \mathfrak{M}_a$ where $\{\mathfrak{M}_a | a \in A\}$ is a collection of $\mathfrak{S}$-irreducible $\mathfrak{S}$-submodules of $\mathfrak{M}$. Let $\{\mathfrak{M}_{a'} | a' \in A'\}$ be a maximal set of k-independent $\mathfrak{S}$-submodules from the collection $\{\mathfrak{M}_a | a \in A\}$. This exists by Zorn's lemma. Then $\mathfrak{M} = \sum_{a' \in A'} \oplus \mathfrak{M}_{a'}$, for otherwise $\mathfrak{M}_b \not\subset \sum \oplus \mathfrak{M}_{a'}$ for some $b \in A$. But then $\mathfrak{M}_b \cap \sum \oplus \mathfrak{M}_{a'} \neq \mathfrak{M}_b$, and $\mathfrak{M}_b \cap \sum \oplus \mathfrak{M}_{a'} = \{0\}$ since $\mathfrak{M}_b$ is $\mathfrak{S}$-irreducible, contradicting the choice of $\{\mathfrak{M}_{a'} | a' \in A'\}$ as a maximal set of k-independent $\mathfrak{S}$-submodules from the collection $\{\mathfrak{M}_a | a \in A\}$. Thus, $2 \Rightarrow 3$. Suppose finally that $\mathfrak{M} = \sum \oplus \mathfrak{M}_a$ where $\{\mathfrak{M}_a | a \in A\}$ is a collection of $\mathfrak{S}$-irreducible $\mathfrak{S}$-submodules of $\mathfrak{M}$. Let $\mathfrak{N}$ be an $\mathfrak{S}$-submodule of $\mathfrak{M}$. By Zorn's lemma, there is a maximal $\mathfrak{S}$-submodule $\mathfrak{N}'$ of $\mathfrak{M}$ such that $\mathfrak{N} \cap \mathfrak{N}' = \{0\}$. For each $a \in A$, $(\mathfrak{N} \oplus \mathfrak{N}') \cap \mathfrak{M}_a$ is either $\{0\}$ or $\mathfrak{M}_a$. In the first case, $\mathfrak{N} \oplus \mathfrak{N}' \oplus \mathfrak{M}_a$ is direct and $\mathfrak{N} \cap (\mathfrak{N}' \oplus \mathfrak{M}_a) = \{0\}$, so that $\mathfrak{M}_a = \{0\}$ by the maximality of $\mathfrak{N}'$. In the second, $\mathfrak{M}_a \subset \mathfrak{N} \oplus \mathfrak{N}'$. Thus, $\mathfrak{N} \oplus \mathfrak{N}'$ contains $\mathfrak{M}_a$ for all $a \in A$, and $\mathfrak{M} = \mathfrak{N} \oplus \mathfrak{N}'$. Thus, $\mathfrak{M}$ is $\mathfrak{S}$-completely reducible and $3 \Rightarrow 1$.

1.2.7 Definition

The *socle* of $\mathfrak{M}$ is Socle $\mathfrak{M} = \sum_{a \in A} \mathfrak{M}_a$ where $\{\mathfrak{M}_a \mid a \in A\}$ is the collection of all $\mathfrak{S}$-irreducible $\mathfrak{S}$-submodules of $\mathfrak{M}$.

By the preceding theorem, Socle $\mathfrak{M}$ is the unique maximal $\mathfrak{S}$-completely reducible $\mathfrak{S}$-submodule of $\mathfrak{M}$.

1.3 Ascent and Descent

In this section we discuss a process of passing from an $\mathfrak{S}$-module $\mathfrak{M}$ over k to an $\mathfrak{S}$-module $\mathfrak{M}'$ over an extension field k' of k. We call this process "ascent." We then discuss "descent," which is a process through which one passes from an $\mathfrak{S}'$-module $\mathfrak{M}'$ over k' to an $\mathfrak{S}$-module $\mathfrak{M}$ over a subfield k of k' such that, roughly speaking, $\mathfrak{M}'$ is obtained from $\mathfrak{M}$ by ascent to k'.

We begin by discussing such processes for vector spaces. For this, let V be a vector space over k and let k' be an extension field of k.

1.3.1 Definition

Let $V_{k'}$ be $k' \otimes_k V$ as a vector space over k', the scalar product being such that $\alpha(\beta \otimes v) = (\alpha\beta) \otimes v$ for $\alpha, \beta \in k'$, $v \in V$.

It is convenient to identify the points of V with points of $V_{k'}$ by means of the k-linear injection $v \to 1 \otimes v$. Then V is a k-subspace of $V_{k'}$, and a k-basis for V is a k'-basis for $V_{k'}$. And for W a subspace of V, $W_{k'}$ is the k'-span of W in $V_{k'}$ where it is understood that $k' \otimes_k W$ is imbedded in $k' \otimes_k V$ in the canonical way.

We also identify the points of $\operatorname{Hom}_k(V, W)$ (W being a second vector space over k) with points of $\operatorname{Hom}_{k'}(V_{k'}, W_{k'})$ via the k-linear injection $T \to \mathrm{id}_{k'} \otimes_k T$.

The process just described takes us from vector spaces and linear transformations over k to vector spaces and linear transformations over an extension field k' of k.

We now carry the discussion over to $\mathfrak{S}$-modules. For this, let $\mathfrak{M}$ be an $\mathfrak{S}$-module over k and let k' be an extension field of k.

1.3.2 Definition

$\mathfrak{M}_{k'} = k' \otimes_k \mathfrak{M}$ as an $\mathfrak{S}$-module over k', the underlying mapping $\mathfrak{M}_{k'} \otimes \mathfrak{S} \to \mathfrak{M}_{k'}$ being given by

$$(\sum \gamma_i \otimes m_i, s) \mapsto \sum \gamma_i \otimes (m_i s)$$

for $m_i \in \mathfrak{M}$, $\gamma_i \in k'$, $s \in \mathfrak{S}$.

As for vector spaces, we identify $\mathfrak{M}$ with a k-subspace of $\mathfrak{M}_{k'}$. For $T \in \mathfrak{S}$, recall that $T_{\mathfrak{M}}$ is the linear transformation of $\mathfrak{M}$ induced by T.

Let $T_{\mathfrak{M}_{k'}} = \mathrm{id}_{k'} \otimes T_{\mathfrak{M}}$. We identify $T_{\mathfrak{M}}$ and $T_{\mathfrak{M}_{k'}}$, according to our convention for vector spaces that $\mathrm{Hom}_k(V, W) \subset \mathrm{Hom}_{k'}(V_{k'}, W_{k'})$. We similarly identify $\mathrm{Hom}_k^{\mathfrak{S}}(\mathfrak{M}, \mathfrak{N})$ ($\mathfrak{N}$ a second $\mathfrak{S}$-module over k) with a k-subspace of $\mathrm{Hom}_{k'}^{\mathfrak{S}}(\mathfrak{M}_{k'}, \mathfrak{N}_{k'})$ according to the same convention.

We now introduce the ideas involved in descent, and begin with a vector space V' over a field k' and a subfield k of k'.

1.3.3 Definition

A *k-form* of V' is a k-subspace V of V' such that a k-basis for V is a k'-basis for V'.

An example of this is $V' = k'^n$ and $V = k^n$. Another is that $1 \otimes_k V$ is a k-form of $V_{k'}$ for any vector space V over k.

1.3.4 Proposition

Let V be a k-subspace of V'. Then the following conditions are equivalent.

1. V is a k-form of V';
2. V' is the k'-span of V and every k-free subset of V is k'-free in V';
3. the k'-linear mapping $V_{k'} \to V'$ such that $\gamma \otimes v \mapsto \gamma v$ for $\gamma \in k'$, $v \in V$ is a k'-isomorphism.

PROOF. It is obvious that $1 \Rightarrow 2$. Since $1 \otimes_k V$ is a k-form of $V_{k'}$, $3 \Rightarrow 1$. It remains to show that $2 \Rightarrow 3$, so we assume 2. The mapping $V_{k'} \to V'$ described in 3 is surjective since the k'-span of V is V'. To show that it is injective, let $x \in V_{k'}$ and write $x = \sum_1^n \gamma_i \otimes v_i$, where $\{v_1, \ldots, v_n\}$ is k-free in V and $\gamma_i \in k'$ $(1 \leq i \leq n)$. The image $\sum_1^n \gamma_i v_i$ of x in V' is 0 iff $\gamma_i = 0$ for $1 \leq i \leq n$, since $\{v_1, \ldots, v_n\}$ is k'-free in V' by 2. Thus, $\sum_1^n \gamma_i v_i = 0$ iff $\sum_1^n \gamma_i \otimes v_i = 0$, and the mapping is injective.

We now fix a k-form V of V', and introduce notions of k-rationality and k-definition relative to this k-form V.

1.3.5 Definition

For $W' \subset V'$, $W'_k = W' \cap V$. In particular, $V'_k = V$.

1.3.6 Definition
A point x of V' is *k-rational* if $x \in V'_k$.

1.3.7 Definition
A subspace W' of V' is *defined over k* if W'_k is a k-form of W'.

Obviously, a subspace W' of V' is defined over k iff W' is the k'-span of W'_k.

1.3.8 Proposition
Suppose that V' is finite dimensional over k' and let $\{W'_a | a \in A\}$ be a collection of subspaces of V' over k' defined over k. Then $\bigcap W'_a$ and $\sum W'_a$ are defined over k.

PROOF. Obviously $V \cap \sum W'_a$ spans $\sum W'_a$ over k', since $V \cap W'_a$ spans W'_a over k' for $a \in A$. Thus, $\sum W'_a$ is defined over k. In showing that $\bigcap W'_a$ is defined over k, we may assume without loss of generality that A is finite, since we have the descending chain condition on k'-subspaces of V'. It therefore suffices to show that if W'_1, W'_2 are k'-subspaces of V' defined over k, then so is $W'_1 \cap W'_2$. Now $W'_1 + W'_2$ is defined over k, so we may assume that $V' = W'_1 + W'_2$ and $V = W_1 + W_2$ where $W_i = W'_i \cap V$ $(i = 1, 2)$. Now $\dim_k(W_1 \cap W_2) = \dim_{k'}(W'_1 \cap W'_2)$ since $\dim V = \dim W_1 + \dim W_2 - \dim (W_1 \cap W_2)$ and $\dim V' = \dim W'_1 + \dim W'_2 - \dim (W'_1 \cap W'_2)$. It follows that $W_1 \cap W_2$ is a k-form of $W'_1 \cap W'_2$ and $W'_1 \cap W'_2$ is defined over k.

We next let V'_1, V'_2 be vector spaces over k', V'_{ik} a fixed k-form of V'_i $(i = 1, 2)$. The following discussion is relative to these fixed k-forms.

1.3.9 Definition
An element $T \in \mathrm{Hom}_{k'}(V'_1, V'_2)$ is *defined over k* if $V'_{1_k} T \subset V'_{2k}$.

1.3.10 Proposition
Let V'_i be finite dimensional over k' $(i = 1, 2)$ and let $T \in \mathrm{Hom}_{k'}(V'_1, V'_2)$ be defined over k. Then Kernel T and Image T are defined over k.

PROOF. $V_1'T$ is defined over k, as the k'-span of $V'_{1_k}T$. Now $\dim_{k'}$ Kernel $T = \dim_{k'} V_1' - \dim_{k'} V_1'T = \dim_k V'_{1_k} - \dim_k V'_{1_k}T = \dim_k$ Kernel $T|_{V'_{1_k}}$. Thus, Kernel $T \cap V'_{1_k}$ spans Kernel T over k' and Kernel T is defined over k.

1.3.11 Definition

$\mathrm{Hom}_{k'}(V_1', V_2')_k = \{T \in \mathrm{Hom}_{k'}(V_1', V_2') \mid T \text{ is defined over } k\}$.

1.3.12 Proposition

Let V_i' be finite dimensional over k' ($i = 1, 2$). Then $\mathrm{Hom}_{k'}(V_1', V_2')_k$ is a k-form of $\mathrm{Hom}_{k'}(V_1', V_2')$ and the mapping $\mathrm{Hom}_{k'}(V_1', V_2')_k \to \mathrm{Hom}_k(V'_{1_k}, V'_{2_k})$ given by $T \mapsto T|_{V'_{1_k}}$ is a k-isomorphism.

PROOF. The second assertion follows immediately from the fact that a k-basis for V'_{1_k} (respectively V'_{2_k}) is a k'-basis for V_1' (respectively V_2'). The first assertion follows from the second upon comparison of dimensions.

We now carry the discussion over to $\mathfrak{S}$-modules. Thus, let $\mathfrak{M}'$ be an $\mathfrak{S}'$-module over k', k a subfield of k'.

1.3.13 Definition

Let $\mathfrak{M}$ be a k-form of the vector space $\mathfrak{M}'$ and let $\mathfrak{S}$ be a subset of $\mathfrak{S}'$ such that $\mathfrak{M}s \subset \mathfrak{M}$ for $s \in \mathfrak{S}$. Then the $\mathfrak{S}$-module $\mathfrak{M}$ over k defined by taking $\mathfrak{M} \times \mathfrak{S} \to \mathfrak{M}$ to be $(m, s) \mapsto ms$ for $m \in \mathfrak{M}$, $s \in \mathfrak{S}$ is a *k-form* of the $\mathfrak{S}'$-module $\mathfrak{M}'$ if the set $\{T_{\mathfrak{M}'} \mid T \in \mathfrak{S}'\}$ of linear transformations of $\mathfrak{M}'$ induced by $\mathfrak{S}'$ is contained in the k'-span of the set $\{T_{\mathfrak{M}'} \mid T \in \mathfrak{S}\}$ of linear transformations of $\mathfrak{M}'$ induced by the elements of $\mathfrak{S}$. An element s of $\mathfrak{S}'$ is *defined over* k relative to a k-form $(\mathfrak{M}, \mathfrak{S})$ of $(\mathfrak{M}', \mathfrak{S}')$ if $\mathfrak{M}s \subset \mathfrak{M}$. The set of elements of $\mathfrak{S}'$ defined over k is denoted $\mathfrak{S}'_k$.

An example of this is $\mathfrak{M}' = \mathfrak{M}_{k'}$, $\mathfrak{S}' = \mathfrak{S}$, and $\mathfrak{M}' \times \mathfrak{S}' \to \mathfrak{M}'$ given by $(\sum \gamma_i \otimes m_i, s) \mapsto \sum \gamma_i \otimes m_i s$ for $\gamma_i \in k'$, $m_i \in \mathfrak{M}$, $s \in \mathfrak{S}$, where $\mathfrak{M}$ is any $\mathfrak{S}$-module over k and k' is any extension field of k. Another example is as follows. Let $\mathfrak{M}$ be a k-form of a finite-dimensional vector space $\mathfrak{M}'$ over k', $\mathfrak{S}' = \mathrm{Hom}_{k'}(\mathfrak{M}', \mathfrak{M}')$. Then $(\mathfrak{M}', \mathfrak{S}')$ is a k'-module and $(\mathfrak{M}, \mathfrak{S})$ is a k-form of $(\mathfrak{M}', \mathfrak{S}')$, where $\mathfrak{S} = \mathrm{Hom}_{k'}(\mathfrak{M}', \mathfrak{M}')_k$, by 1.3.12.

We now let $(\mathfrak{M}, \mathfrak{S}_{\mathfrak{M}})$, $(\mathfrak{N}, \mathfrak{S}_{\mathfrak{N}})$ be k-forms of $(\mathfrak{M}', \mathfrak{S}')$, $(\mathfrak{N}', \mathfrak{S}')$, respectively, and suppose that $\mathfrak{S}_{\mathfrak{M}} \subset \mathfrak{S}_{\mathfrak{N}}$.

1.3.14 Definition
An $\mathfrak{S}'$-homomorphism $f: \mathfrak{M}' \to \mathfrak{N}'$ of $\mathfrak{S}'$-modules $\mathfrak{M}', \mathfrak{N}'$ over k' is *defined over* k relative to the k-forms $\mathfrak{M}$, $\mathfrak{N}$ if $f(\mathfrak{M}) \subset \mathfrak{N}$. The set of $\mathfrak{S}'$-homomorphisms from $\mathfrak{M}'$ to $\mathfrak{N}'$ defined over k is denoted $\operatorname{Hom}_{k'}^{\mathfrak{S}'}(\mathfrak{M}', \mathfrak{N}')_k$.

1.3.15 Proposition
If $\mathfrak{M}', \mathfrak{N}'$ are finite-dimensional $\mathfrak{S}'$-modules over k', then $\operatorname{Hom}_{k'}^{\mathfrak{S}'}(\mathfrak{M}', \mathfrak{N}')_k$ is a k-form of $\operatorname{Hom}_{k'}^{\mathfrak{S}'}(\mathfrak{M}', \mathfrak{N}')$.

PROOF. We can express an element of $\operatorname{Hom}_{k'}^{\mathfrak{S}'}(\mathfrak{M}', \mathfrak{N}')$ as $f = \sum_1^n \gamma_i f_i$ where $\{\gamma_1, \ldots, \gamma_n\}$ is k-free in k' and $\{f_1, \ldots, f_n\} \subset \operatorname{Hom}_{k'}(\mathfrak{M}', \mathfrak{N}')_k$, by 1.3.12. Since f commutes with the elements of $\mathfrak{S}_{\mathfrak{M}}$, so do the f_i. Here one needs the assumption that $\mathfrak{S}_{\mathfrak{M}} \subset \mathfrak{S}_{\mathfrak{N}}$. Thus, the f_i commute with the elements of $\mathfrak{S}'$ since the transformations induced by $\mathfrak{S}'$ on $\mathfrak{M}'$ are in the k'-span of the set of transformations induced by $\mathfrak{S}_{\mathfrak{M}}$ on $\mathfrak{M}'$.

1.4 Jordan Decomposition
Let $\mathfrak{S}$ be a set, $\mathfrak{M}$ an $\mathfrak{S}$-module over k. Recall that for $T \in \mathfrak{S}$, $T_{\mathfrak{M}}$ is the linear transformation of $\mathfrak{M}$ induced by T. We let I be the identity transformation $\mathrm{id}_{\mathfrak{M}}$ of $\mathfrak{M}$.

1.4.1 Definition
Let $T \in \mathfrak{S}$ and $\alpha \in k$. Then $\mathfrak{M}_\alpha(T) = \{m \in \mathfrak{M} \mid m(T_{\mathfrak{M}} - \alpha I)^n = 0$ for some positive integer $n\}$.

1.4.2 Proposition
Let k be algebraically closed and M finite dimensional over k. Then $\mathfrak{M} = \sum_{\alpha \in k} \oplus \mathfrak{M}_\alpha(T)$ for $T \in \mathfrak{S}$.

PROOF. Let $f(X)$ be the minimum polynomial of $T_{\mathfrak{M}}$. Thus, $f(X)$ is a monic polynomial with coefficients in k, $f(T_{\mathfrak{M}}) = 0$ and degree $f(X)$ is minimal with these properties. Let $f(X) = \prod(X - \alpha_i)^{e_i}$ and choose $g_i(X) \in k[X]$ such that $\sum g_i(X)h_i(X) = 1$ where $h_i(X) = \dfrac{f(X)}{(X - \alpha_i)^{e_i}}$. Here, we are supposing that the α_i are distinct. Hence the greatest common factor of the $h_i(X)$ is 1, so that such $g_i(X)$ do exist. Now $\mathfrak{M} = \sum \mathfrak{M}_i$ where $\mathfrak{M}_i = \mathfrak{M}h_i(T_{\mathfrak{M}})$, since $m = \sum(mg_i(T_{\mathfrak{M}}))h_i(T_{\mathfrak{M}})$ for $m \in \mathfrak{M}$. And $\mathfrak{M}_i \subset \mathfrak{M}_{\alpha_i}(T)$, since $\mathfrak{M}_i(T_{\mathfrak{M}} - \alpha_i)^{e_i} = \mathfrak{M}h_i(T_{\mathfrak{M}})(T_{\mathfrak{M}} - \alpha_i)^{e_i} = \mathfrak{M}f(T_{\mathfrak{M}}) = \{0\}$. Thus, $\mathfrak{M} = \sum \mathfrak{M}_{\alpha_i}(T)$ and it remains only to show that this sum is direct. Thus, let $m \in \mathfrak{M}_{\alpha_i}(T) \cap \sum_{i \neq j} \mathfrak{M}_{\alpha_j}(T)$. Then the ideal $I = \{h(X) \in k[X] \mid mh(T_{\mathfrak{M}}) = 0\}$ contains $(X - \alpha_i)^{f_i}$ and $\prod_{j \neq i} (X - \alpha_j)^{f_j}$ for suitable f_i, f_j. Since the latter are relatively prime, $I = k[X]$ and $1 \in I$. Thus, $m = m1 = 0$ and the sum must be direct.

1.4.3 Definition

We say that T (or $T_{\mathfrak{M}}$) *splits over k* if $\mathfrak{M} = \sum_{\alpha \in k} \mathfrak{M}_\alpha(T)$ or, equivalently, if the minimum polynomial $f(X)$ of $T_{\mathfrak{M}}$ can be factored over k into linear factors $f(X) = \prod(X - \alpha_i)^{e_i}$.

1.4.4 Proposition

If $f(X)$ is the minimum polynomial of $T_{\mathfrak{M}}$ over k and $\alpha_1, \ldots, \alpha_n$ are the roots of $f(X)$ in some algebraically closed extension field of k, then T (or $T_{\mathfrak{M}}$) splits over the finite-dimensional extension field k' of k generated by $\alpha_1, \ldots, \alpha_n$.

PROOF. $f(X) = \prod(X - \alpha_i)^{e_i}$ in $k'[X]$ so that the minimum polynomial of T over k', which is in fact $f(X)$, must have a similar factorization into linear factors.

1.4.5 Definition

Let T be a linear transformation of a finite-dimensional vector space V over k and let T be split over k. Then the *semisimple part* of T is the linear transformation T_s such that $T_s|_{V_\alpha(T)} = \alpha \cdot \mathrm{id}_{V_\alpha(T)}$ for $\alpha \in k$. The ***nilpotent part*** of T is $T_n = T - T_s$. The decomposition $T = T_s + T_n$ is called the *Jordan decomposition* of T. If $T = T_s$, T is *semisimple*. If $T = T_n$, T is *nilpotent*.

Note that T_n and T_s commute, since they both stabilize each $V_\alpha(T)$ and since $T_s|_{V\alpha(T)}$ is a scalar for all α. It follows that T is nilpotent iff $T^n = 0$ for some positive integer n.

1.4.6 Proposition

Let T be a linear transformation of a finite-dimensional vector space V over k and assume that T is split over k. Let $T = S + N$, where S is semisimple, N is nilpotent, and $SN = NS$. Then $S = T_s$ and $N = T_n$.

PROOF. S and N commute with T, so they stabilize the $V_\alpha(T)$. On $V_\alpha(T)$, $S = T - N$ has only one eigenvalue, namely α. Since S is semisimple, this means that $S|_{V\alpha(T)} = \alpha\, \mathrm{id}_{V\alpha(T)}$. Thus, $S = T_s$ since they agree on all $V_a(T)$. Now $N = T - S = T - T_s = T_n$.

It can be shown that T_s and T_n are polynomials in T. The following proposition is a weak version of this which is adequate for our purposes.

1.4.7 Proposition

Let A be a linear transformation of a finite-dimensional vector space V over k which commutes with a split linear transformation T of V over k. Then A commutes with T_s and T_n. Any T-stable subspace of V is stable under T_s and T_n.

PROOF. The transformation A stabilizes the $V_\alpha(T)$, hence commutes with T_s, hence with $T_n = T - T_s$. If W is T-stable, then $W = \sum \oplus W_\alpha(T)$. Since $W_\alpha(T) \subset V_\alpha(T)$ and $T_s|_{V\alpha(T)} = \alpha\, \mathrm{id}_{V\alpha(T)}$, T_s stabilizes W. Thus, $T_n = T - T_s$ stabilizes W.

1.4.8 Proposition

Let T be split over k. Define $\mathrm{ad}\, T \in \mathrm{Hom}\,(\mathrm{Hom}\, V)$ by $S\, \mathrm{ad}\, T = ST\text{–}TS$ for $S \in \mathrm{Hom}\, V$. Then $\mathrm{ad}\, T$ is split over k, and $(\mathrm{ad}\, T)_s = \mathrm{ad}\, T_s$ and $(\mathrm{ad}\, T)_n = \mathrm{ad}\, T_n$.

PROOF. Since $T_s T_n = T_n T_s$, we have $\mathrm{ad}\, T_s\, \mathrm{ad}\, T_n = \mathrm{ad}\, T_n\, \mathrm{ad}\, T_s$. Thus, it suffices to show that $\mathrm{ad}\, T_s$ is split over k and semisimple, and $\mathrm{ad}\, T_n$

nilpotent. For this, let $e_1, \ldots, e_n$ be a basis for V consisting of eigenvectors for T_s and let $e_iT_s = \alpha_i e_i$ for $1 \leq i \leq n$. Let $\{E_{ij} \mid 1 \leq i, j \leq n\}$ be the basis for $\mathrm{Hom}_k V$ such that $e_kT_{ij} = \delta_{ki}e_j$ where δ_{ki} is 0 for $k \neq i$ and 1 for $k = i$. Then E_{ij} ad $T_s = (\alpha_j - \alpha_i)E_{ij}$ for $1 \leq i, j \leq n$ and ad T_s is split over k and semisimple. On the other hand, $Y(\mathrm{ad}\, T_n)^m$ is a sum of terms $T_n^rYT_n^{m-r}$, so that $T_n^q = 0$ implies that $(\mathrm{ad}\, T_n)^{2q} = 0$, and ad T_n is nilpotent.

We now generalize the foregoing discussion, and begin with a review of some Galois theory. We refer the reader to [14] for details. Thus, let k' be an algebraic field extension of k and let $\bar{k}$ be an algebraic closure of k containing k'. Then k' is *Galois* over k if k' is G-stable and $k'^G = \{\chi \in k' \mid \chi g = \chi \text{ for all } g \in G\}$ is equal to k for $G = \mathrm{Aut}\, \bar{k}/k$, the group of k-linear automorphisms of $\bar{k}$. We say that k is *perfect* if the algebraic closure of k is Galois over k. Every field of characteristic 0 is perfect, and every finite field is perfect. A polynomial $f(X) \in k[X]$ is *separable* over k if there is an algebraic Galois extension k' of k such that $f(X)$ splits over k'.

1.4.9 Definition

Let $T \in \mathfrak{S}$. Then T (or $T_{\mathfrak{M}}$) is *separable* over k if the minimum polynomial of $T_{\mathfrak{M}}$ over k is separable over k.

If k is perfect (e.g., if k is finite or of characteristic 0), then every $T \in \mathfrak{S}$ is separable over k, for we may take k' to be the algebraic closure of k.

1.4.10 Proposition

Let V be a finite-dimensional vector space over k, and let T be a separable element of $\mathfrak{S} = \mathrm{Hom}_k V$. Let k' be an algebraic Galois extension of k such that the element $T' = \mathrm{id}_{k'} \otimes T$ of $\mathrm{Hom}_{k'} V_{k'}$ is split over k'. Then the semisimple and nilpotent parts $T'_s = T'_n$ of T' are defined over k.

PROOF. We introduce the usual imbeddings $V \subset V_{k'} \subset V_{\bar{k}}$ and $\mathrm{Hom}_k V \subset \mathrm{Hom}_{k'} V' \subset \mathrm{Hom}_{\bar{k}} V_{\bar{k}}$, and let G act on $\mathrm{Hom}_{\bar{k}} V_{\bar{k}}$ by defining A^g for $A \in \mathrm{Hom}_{\bar{k}} V_{\bar{k}}$ by the commutative diagram

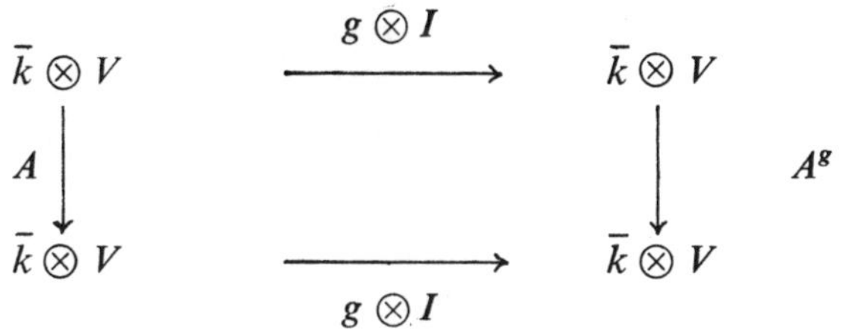

Figure 1.

for $g \in G$. Since k' is G-stable, $\mathrm{Hom}_{k'} V_{k'}$ is a G-stable subset of $\mathrm{Hom}_{\bar{k}} V_{\bar{k}}$. Now T' is defined over k and k is fixed pointwise by the elements g of G. Thus, $T' = T'^g = T_s'^g + T_n'^g$ for g in G. Since $T_s', T_n' \in \mathrm{Hom}_{k'} V_{k'}$, $T_s'^g, T_n'^g \in \mathrm{Hom}_{k'} V_{k'}$. Thus, by the unicity given in 1.4.6, we have $T_s' = T_s'^g$ and $T_n' = T_n'^g$ for all $g \in G$. But $k = \{\chi \in k' \mid \chi g = \chi \text{ for all } g \in G\}$, and it follows that $\mathrm{Hom}_k V = \{A \in \mathrm{Hom}_{k'} V_{k'} | A^g = A \text{ for all } g \in G\}$. Thus, $T_s', T_n' \in \mathrm{Hom}_k V$. That is, T_s' and T_n' are defined over k.

1.4.11 Definition

Let T be a separable element of $\mathrm{Hom}_k V$, where V is a finite-dimensional vector space over k. Then $T_s = T_s'|_V$ and $T_n = T_n'|_V$ where T' and k' are as in 1.4.10. If $T = T_s$, T is *semisimple*. If $T = T_n$, T is *nilpotent*.

One shows that the choice of k' does not affect this definition, by an application of 1.4.6. One also proves the following generalizations of earlier observations. The details are straightforward and are left to the reader.

1.4.12 Proposition

Let T be a separable element of $\mathfrak{S} = \mathrm{Hom}_k V$, where V is a finite-dimensional vector space over k. Then

1. T_s and T_n commute;
2. T is nilpotent iff $T^n = 0$ for some $n > 0$;
3. if $T = S + N$, where S is semisimple, N is nilpotent, and $SN = NS$, then $S = T_s$ and $N = T_n$;
4. if $A \in \mathrm{Hom}_k V$ commutes with T, then A commutes with T_s and T_n; any T-stable subspace of V is stable under T_s and T_n;
5. ad T is separable over k and $(\mathrm{ad}\, T)_s = \mathrm{ad}\, T_s$, $(\mathrm{ad}\, T)_n = \mathrm{ad}\, T_n$.

1.5 Fitting Decomposition

Let $\mathfrak{M}$ be a finite-dimensional $\mathfrak{S}$-module over a field k, and let T be an element of $\mathfrak{S}$. Note that $\mathfrak{M}_0(T) = \bigcup_{i=0}^{\infty} \text{Kernel } (T_{\mathfrak{M}})^i$, by 1.4.1.

1.5.1 Definition

$\mathfrak{M}_*(T) = \bigcap_{i=0}^{\infty} \text{Image } (T_{\mathfrak{M}})^i$. We now restate 1.4.3.

1.5.2 Proposition

Let T be split over k. Then $\mathfrak{M}_*(T) = \sum_{\alpha \in k^*} \mathfrak{M}_\alpha(T)$ where $k^* = k - \{0\}$, and $\mathfrak{M} = \mathfrak{M}_0(T) \oplus \mathfrak{M}_*(T)$.

The decomposition $\mathfrak{M} = \mathfrak{M}_0(T) \oplus \mathfrak{M}_*(T)$ is the *Fitting decomposition* of $\mathfrak{M}$ with respect to T. The spaces $\mathfrak{M}_0(T)$, $\mathfrak{M}_*(T)$ are T-stable and are called the zero and one components of $\mathfrak{M}$ with respect to T. The Fitting decomposition holds also when T is not split. Before proving this, we broaden our point of view.

1.5.3 Definition

Let $a: \mathfrak{S} \to k$ be a function. Then $\mathfrak{M}_a(\mathfrak{S}) = \bigcap_{T \in \mathfrak{S}} \mathfrak{M}_{a(T)}(T)$.

The zero function $\mathfrak{S} \to \{0\}$ is denoted 0, so that $\mathfrak{M}_0(\mathfrak{S}) = \bigcap_{T \in \mathfrak{S}} \mathfrak{M}_0(T)$. We let $\mathfrak{M}_*(\mathfrak{S}) = \Sigma_{T \in \mathfrak{S}} \mathfrak{M}_*(T)$. The spaces $\mathfrak{M}_0(\mathfrak{S})$, $\mathfrak{M}_*(\mathfrak{S})$ are the *Fitting components* of $\mathfrak{M}$ with respect to $\mathfrak{S}$.

1.5.4 Proposition

Let $\mathfrak{M}'$ be a finite-dimensional $\mathfrak{S}'$-module over k', k a subfield of k', $(\mathfrak{M}, \mathfrak{S})$ a k-form of the $\mathfrak{S}'$-module $(\mathfrak{M}', \mathfrak{S}')$. Then $\mathfrak{M}'_0(\mathfrak{S}')$ and $\mathfrak{M}'_*(\mathfrak{S}')$ are defined over k, $\mathfrak{M}_0(\mathfrak{S})$ is a k-form of $\mathfrak{M}'_0(\mathfrak{S}')$ and $\mathfrak{M}_*(\mathfrak{S})$ a k-form of $\mathfrak{M}'_*(\mathfrak{S}')$. Moreover if $a: \mathfrak{S} \to k$ is a k-valued function, then $\mathfrak{M}'_a(\mathfrak{S})$ is defined over k and $\mathfrak{M}_a(\mathfrak{S})$ is a k-form of $\mathfrak{M}'_a(\mathfrak{S})$.

PROOF. Throughout the proof, we freely use the equalities $\mathfrak{M}'_0(\mathfrak{S}') = \mathfrak{M}'_0(\mathfrak{S})$ and $\mathfrak{M}'_*(\mathfrak{S}') = \mathfrak{M}'_*(\mathfrak{S})$. These follow from the assumption for k-forms that $\{T_{\mathfrak{M}'} \mid T \in \mathfrak{S}'\}$ is contained in the k'-span of $\{T_{\mathfrak{M}'} \mid T \in \mathfrak{S}\}$. We begin by showing that $\mathfrak{M}'_0(\mathfrak{S})$, $\mathfrak{M}'_a(\mathfrak{S})$, $\mathfrak{M}'_*(\mathfrak{S})$ are defined over k, $a: \mathfrak{S} \to k$ being a given function. For $T \in \mathfrak{S}$, $\mathfrak{M}'_0(T) = \sum_{i=0}^{\infty} \text{Kernel } T^i_{\mathfrak{M}}$ is defined over k, by 1.3.8 and 1.3.10. Similarly, $\mathfrak{M}'_{a(T)}(T) = \mathfrak{M}'_0(T - a(T)I)$ is defined over k

for $T \in \mathfrak{S}$. Thus, $\mathfrak{M}'_0(\mathfrak{S}) = \bigcap_{T \in \mathfrak{S}} \mathfrak{M}'_0(T)$ and $\mathfrak{M}'_a(\mathfrak{S}) = \bigcap_{T \in \mathfrak{S}} \mathfrak{M}'_{a(T)}(T)$ are defined over k, by 1.3.8. Finally, $\mathfrak{M}'_*(T) = \bigcap_{i=0}^{\infty} \text{Image } T^i_{\mathfrak{M}}$ is defined over k for $T \in \mathfrak{S}$, by 1.3.8 and 1.3.10, so that $\mathfrak{M}'_*(\mathfrak{S}) = \sum_{T \in \mathfrak{S}} \mathfrak{M}'_*(T)$ is defined over k, by 1.3.8.

We next note that $\mathfrak{M}_a(\mathfrak{S})$ is a k-form of $\mathfrak{M}'_a(\mathfrak{S})$, for since $\mathfrak{M}'_a(\mathfrak{S})$ is defined over k, $\mathfrak{M}_a(\mathfrak{S}) = \mathfrak{M} \cap \mathfrak{M}'_a(\mathfrak{S})$ is a k-form of $\mathfrak{M}'_a(\mathfrak{S})$. Taking $a = 0$, $\mathfrak{M}_0(\mathfrak{S})$ is a k-form of $\mathfrak{M}'_0(\mathfrak{S})$.

Finally, we show that $\mathfrak{M}_*(\mathfrak{S})$ is a k-form of $\mathfrak{M}'_*(\mathfrak{S})$. Since $\mathfrak{M}'_*(\mathfrak{S})$ is defined over k, $\mathfrak{W} = \mathfrak{M} \cap \mathfrak{M}'_*(\mathfrak{S})$ is a k-form of $\mathfrak{M}'_*(\mathfrak{S})$. Since $\mathfrak{W}$ is a k-form of $\mathfrak{M}'_*(\mathfrak{S})$, the k'-span of $\mathfrak{W}_*(T) = \bigcap_{i=0}^{\infty} \mathfrak{W} T^i_{\mathfrak{M}}$ is $\bigcap_{i=0}^{\infty} \mathfrak{M}'_*(\mathfrak{S}) T^i_{\mathfrak{M}'} = \mathfrak{M}'_*(T)$ for $T \in \mathfrak{S}$. Thus, the k'-span of $\mathfrak{W}_*(\mathfrak{S}) = \sum_{T \in \mathfrak{S}} \mathfrak{W}_*(T)$ is $\sum_{T \in \mathfrak{S}} \mathfrak{M}'_*(T) = \mathfrak{M}'_*(\mathfrak{S})$. Now $\mathfrak{W}_*(\mathfrak{S}) \subset \mathfrak{M}_*(\mathfrak{S})$, so that the k'-span of $\mathfrak{M}_*(\mathfrak{S})$ is $\mathfrak{M}'_*(\mathfrak{S})$, and $\mathfrak{M}_*(\mathfrak{S})$ is a k-form of $\mathfrak{M}'_*(\mathfrak{S})$.

1.5.5 Corollary

Let $\mathfrak{M}$ be a finite-dimensional $\mathfrak{S}$-module over k. Then $\mathfrak{M} = \mathfrak{M}_0(T) \oplus \mathfrak{M}_*(T)$ for $T \in \mathfrak{S}$.

PROOF. Let k' be the algebraic closure of k, $\mathfrak{M}' = \mathfrak{M}_{k'}$. Then $\mathfrak{M}'$ is an $\mathfrak{S}$-module and $\mathfrak{M}' = \mathfrak{M}'_0(T) \oplus \mathfrak{M}'_*(T)$ for $T \in \mathfrak{S}$. But $\mathfrak{M}_0(T)$, $\mathfrak{M}_*(T)$ are k-forms of $\mathfrak{M}'_0(T)$, $\mathfrak{M}'_*(T)$ respectively, and it follows that $\mathfrak{M} = \mathfrak{M}_0(T(\oplus \mathfrak{M}_*(T)$ for $T \in \mathfrak{S}$.

2

Nonassociative Algebras

2.1 Introduction

A Lie algebra is a nonassociative algebra satisfying certain conditions. We introduce here the language of nonassociative algebras. We then give a brief development of some basic material on nonassociative algebras which is used in Chapters 3 and 4. Some of this material is concerned with conditions under which a given nonassociative algebra is a Lie algebra. The remainder of the material is of interest to us only because it applies to Lie algebras and is developed in this section only because it is most naturally formulated in the language of nonassociative algebras.

Most of the material is fairly well known. A possible exception is the theorem in 2.7.5 on translations in a nonassociative algebra, which generalizes a theorem of Hans Zassenhaus on translations in a Lie algebra. Another possible exception is the material in 2.4.17, 2.6.7, and 2.6.8 on characteristic ideals of a nonassociative algebra. The latter material consists of some interesting theorems which are special cases of a recent result of Richard Block [4].

We begin with some elementary definitions and observations.

2.1.1 Definition

A *nonassociative algebra* over k is a vector space A over k together with a product $m: \mathfrak{A} \times \mathfrak{A} \to \mathfrak{A}$, denoted $(x, y) \mapsto xy$, such that $(\alpha x + \beta y)z = \alpha(xz) + \beta(yz)$ and $z(\alpha x + \beta y) = \alpha(zx) + \beta(zy)$ for all $\alpha, \beta \in k$ and $x, y, z \in \mathfrak{A}$.

If V is a vector space over k with basis $e_1, \ldots, e_n$, then V can be given the structure of a nonassociative algebra over k by selecting elements $v_{ij} \in V$ for $1 \leq i, j \leq n$, defining $e_i e_j = v_{ij}$ and extending this linearly to a product on V. Any nonassociative algebraic structure for V can be obtained in this way. Letting $v_{ij} = \Sigma \Gamma_{ij}^k e_k$, we obtain an element (Γ_{ij}^k) of k^{n^3} whose coordinates are called the structure constants of the algebra. These structure constants completely determine the product. Thus, the set of nonassociative algebras with underlying vector space V over k can be identified with k^{n^3}.

We now let $\mathfrak{A}$ be a nonassociative algebra over a field k.

2.1.2 Definition

Let $\mathfrak{B}$, $\mathfrak{C}$ be subsets of $\mathfrak{A}$. Then we let $\mathfrak{B}\mathfrak{C}$ be the subspace of $\mathfrak{A}$ generated by $\{bc \mid b \in \mathfrak{B}, c \in \mathfrak{C}\}$, and $\mathfrak{B}^2 = \mathfrak{B}\mathfrak{B}$.

2.1.3 Definition

A *subalgebra* of $\mathfrak{A}$ is a subspace $\mathfrak{B}$ of $\mathfrak{A}$ such that $\mathfrak{B}^2 \subset \mathfrak{B}$. An *ideal* of $\mathfrak{A}$ is a subalgebra $\mathfrak{B}$ of $\mathfrak{A}$ such that $\mathfrak{A}\mathfrak{B} \subset \mathfrak{B}$ and $\mathfrak{B}\mathfrak{A} \subset \mathfrak{B}$. If $\mathfrak{B}$ is an ideal of $\mathfrak{A}$, the *quotient algebra* $\mathfrak{A}/\mathfrak{B}$ has the product $(x + \mathfrak{B})(y + \mathfrak{B}) = xy + \mathfrak{B}$.

2.1.4 Definition

The *direct sum* of nonassociative algebras $\mathfrak{A}_i$ over k is the vector space $\sum \oplus \mathfrak{A}_i$ with product $(\sum a_i)(\sum b_i) = \sum a_i b_i$. The *tensor product* of nonassociative algebras $\mathfrak{A}_1, \ldots, \mathfrak{A}_n$ over k is the tensor product $\mathfrak{A}_1 \otimes \ldots \otimes \mathfrak{A}_n$ of the vector spaces $\mathfrak{A}_1, \ldots, \mathfrak{A}_n$ together with the bilinear product determined by $(a_1 \otimes \ldots \otimes a_n)(b_1 \otimes \ldots \otimes b_n) = a_1 b_1 \otimes \ldots \otimes a_n b_n$.

2.1.5 Definition

A *homomorphism* of nonassociative algebras $\mathfrak{A}$, $\mathfrak{B}$ over k is a k-linear mapping $f\colon \mathfrak{A} \to \mathfrak{B}$ such that $f(xy) = f(x)f(y)$ for $x, y \in \mathfrak{A}$. Such an f is an *isomorphism* if it is bijective.

We have the following usual homomorphism theorems.

2.1.6 Theorem

Let $f\colon \mathfrak{A} \to \mathfrak{B}$ be a homomorphism of nonassociative algebras over k. Then Kernel f is an ideal of $\mathfrak{A}$, Image f is a subalgebra of $\mathfrak{B}$, and $x +$ Kernel $f \mapsto f(x)$ is an isomorphism from $\mathfrak{A}/$Kernel f to Image f.

2.1.7 Theorem

Let $\mathfrak{B}$ be an ideal of $\mathfrak{A}$, and $\mathfrak{C}$ a subalgebra of $\mathfrak{A}$. Then $\mathfrak{B} + \mathfrak{C}/\mathfrak{B}$ and $\mathfrak{C}/\mathfrak{B} \cap \mathfrak{C}$ are isomorphic through $x + \mathfrak{B} \mapsto x + \mathfrak{B} \cap \mathfrak{C}$ $(x \in \mathfrak{C})$.

2.1.8 Definition

$\mathfrak{A}$ is *commutative* if $xy = yx$ for $x, y \in \mathfrak{A}$.

2.1.9 Definition

$\mathfrak{A}$ is *anticommutative* if $x^2 = 0$ for all $x \in \mathfrak{A}$. Here, x^2 denotes xx.

If $\mathfrak{A}$ is anticommutative, then $xy = -yx$ for all $x, y \in \mathfrak{A}$, since $0 = (x + y)^2 = x^2 + y^2 + xy + yx = xy + yx$.

2.1.10 Definition

$\mathfrak{A}$ is an *associative algebra* if $(xy)z = x(yz)$ for all $x, y, z \in \mathfrak{A}$.

2.1.11 Definition

$\mathfrak{A}$ is a *Lie algebra* if $\mathfrak{A}$ is anticommutative and $(xy)z + (yz)x + (zx)y = 0$ for all $x, y, z \in \mathfrak{A}$.

The Lie algebra identity described above is the *Jacobi identity*.

To show that $\mathfrak{A}$ is associative or Lie, it suffices to verify the associative law or the Jacobi identity and the condition $x^2 = 0$ for elements of some basis of $\mathfrak{A}$.

2.1.12 Definition

If $\mathfrak{A}$ is an associative algebra, we let $[x, y] = xy - yx$ for $x, y \in \mathfrak{A}$ and let $\mathfrak{A}_{\text{Lie}}$ be the nonassociative algebra with underlying vector space $\mathfrak{A}$ and product $(x, y) \mapsto [x, y]$.

The algebra $\mathfrak{A}_{\text{Lie}}$ is a Lie algebra, as we shall see shortly.

If $\mathfrak{A}$ is associative, it is sometimes convenient to indicate that we are working with it as an associative algebra by writing $\mathfrak{A}_{\text{Assoc}}$ in place of $\mathfrak{A}$. Thus, $(\text{Hom}_k V)_{\text{Assoc}}$ is the associative algebra of linear transformations of a vector space V over k, whereas $(\text{Hom}_k V)_{\text{Lie}}$ is the Lie algebra of linear transformations of V. The *associative* product is XY, the *Lie* product $[X, Y] = XY - YX$.

The following theorem theoretically reduces the study of finite-dimensional Lie algebras to the study of subalgebras of $(\text{Hom}_k V)_{\text{Lie}}$, with V a finite-dimensional vector space over a field k. We do not use this theorem, and refer the reader to [13] for the proof.

2.1.13 Theorem (Ado-Iwasawa)

Let $\mathfrak{L}$ be a finite-dimensional Lie algebra over k. Then $\mathfrak{L}$ is isomorphic to a subalgebra of $(\text{Hom}_k V)_{\text{Lie}}$ for some finite-dimensional vector space V over k.

2.2 Ascent and Descent

Let k' be an extension field of k.

2.2.1 Definition

$\mathfrak{A}_{k'}$ is the nonassociative algebra over k' with underlying vector space $k' \otimes_k \mathfrak{A}$ and product $m_{k'} : \mathfrak{A}_{k'} \otimes \mathfrak{A}_{k'} \to \mathfrak{A}_{k'}$ determined by $(\alpha \otimes c)(\beta \otimes d) = \alpha\beta \otimes cd$ for $\alpha, \beta \in k'$, $c, d \in \mathfrak{A}$.

2.2.2 Proposition

$\mathfrak{A}$ is associative (respectively Lie) iff $\mathfrak{A}_{k'}$ is associative (respectively Lie).

PROOF. We may identify $\mathfrak{A}$ and $1 \otimes \mathfrak{A}$ as in 1.3. Now a basis for $\mathfrak{A}$ is a basis for $\mathfrak{A}_{k'}$, and the assertion follows from a remark in 2.1.

2.2.3 Proposition

Let $\mathfrak{B}$, $\mathfrak{C}$ be subspaces of $\mathfrak{A}$. Then $(\mathfrak{B}\mathfrak{C})_{k'} = \mathfrak{B}_{k'}\mathfrak{C}_{k'}$. (We follow the conventions of 1.3.)

PROOF. Let $\{e_i\}$, $\{f_j\}$ be bases for $\mathfrak{B}$, $\mathfrak{C}$ over k. Then $\{e_i\}$, $\{f_j\}$ are bases for $\mathfrak{B}_{k'}$, $\mathfrak{C}_{k'}$ over k'. Now $\{e_i f_j\}$ is a generating set for $\mathfrak{B}_{k'}\mathfrak{C}_{k'}$ over k' as well as for $(\mathfrak{B}\mathfrak{C})_{k'}$ over k'. Thus, $(\mathfrak{B}\mathfrak{C})_{k'} = \mathfrak{B}_{k'}\mathfrak{C}_{k'}$.

2.2.4 Definition

Let $\mathfrak{A}'$ be a nonassociative algebra over k', k a subfield of k', and $\mathfrak{A}$ a nonassociative algebra over k. Then $\mathfrak{A}$ is a *k-form* of $\mathfrak{A}'$ if the underlying vector space of $\mathfrak{A}$ is a k-form of the underlying vector space of $\mathfrak{A}'$ and, for $x, y \in \mathfrak{A}$, $m(x, y) = m'(x, y)$, where m, m' are the product mappings for $\mathfrak{A}$, $\mathfrak{A}'$ respectively.

2.2.5 Proposition

Let $\mathfrak{A}$ be a k-form of $\mathfrak{A}'$, $\mathfrak{A}$ and $\mathfrak{A}'$ being as in 2.2.4. Then $\mathfrak{A}_{k'}$ and $\mathfrak{A}'$ are isomorphic via the mapping determined by $\alpha \otimes x \mapsto \alpha x$ for $\alpha \in k'$, $x \in \mathfrak{A}$.

PROOF. We leave the proof to the reader.

2.3 Translations and the Multiplication Algebra

2.3.1 Definition

For $a \in \mathfrak{A}$, let L_a and R_a be the elements of $\mathrm{Hom}_k\, \mathfrak{A}$ defined by $xL_a = ax$, $xR_a = xa$ for $x \in \mathfrak{A}$. For $\mathfrak{S} \subset \mathfrak{A}$, let $L_{\mathfrak{S}} = \{L_a \mid a \in \mathfrak{S}\}$, $R_{\mathfrak{S}} = \{R_a \mid a \in \mathfrak{S}\}$.

For $a \in \mathfrak{A}$ and $\gamma \in k$, we have $\gamma L_a = L_{\gamma a}$ and $\gamma R_a = R_{\gamma a}$. It follows that the subring $M(\mathfrak{A})$ of $(\mathrm{Hom}_k\, \mathfrak{A})_{\mathrm{Assoc}}$ generated by $L_{\mathfrak{A}} \cup R_{\mathfrak{A}}$ is a k-subalgebra. The identity of $\mathrm{Hom}_k\, \mathfrak{A}$ need not be contained in $M(\mathfrak{A})$.

2.3.2 Definition

$M(\mathfrak{A})$ is the *multiplication algebra* of $\mathfrak{A}$.

2.3.3 Definition

If $\mathfrak{A}$ is a Lie algebra and $x \in \mathfrak{A}$, we let $\mathrm{ad}_{\mathfrak{A}}\, x = R_x$ and call $\mathrm{ad}_{\mathfrak{A}}\, x$ the *adjoint* of x in $\mathfrak{A}$. We sometimes write $\mathrm{ad}\, x = \mathrm{ad}_{\mathfrak{A}}\, x$. For $\mathfrak{B} \subset \mathfrak{A}$, we let $\mathrm{ad}\, \mathfrak{B}$ or $\mathrm{ad}_{\mathfrak{A}}\, \mathfrak{B}$ denote $\{\mathrm{ad}_{\mathfrak{A}}\, x \mid x \in \mathfrak{B}\}$.

2.3.4 Definition

If $\mathfrak{A}$ is an associative algebra and $x \in \mathfrak{A}$, we let $\mathrm{ad}\, x$ or $\mathrm{ad}_{\mathfrak{A}}\, x$ denote the adjoint of x in $\mathfrak{A}_{\mathrm{Lie}}$, that is, $y\, \mathrm{ad}\, x = xy - yx$ for $y \in \mathfrak{A}$.

We often regard $\mathfrak{A}$ as an $M(\mathfrak{A})$-module in the obvious way. The $M(\mathfrak{A})$-submodules are then the ideals of $\mathfrak{A}$.

2.4 Derivations

2.4.1 Definition

A *derivation* of $\mathfrak{A}$ is a linear transformation D of $\mathfrak{A}$ such that $(xy)D = (xD)y + x(yD)$ for $x, y \in \mathfrak{A}$. The set of derivations of $\mathfrak{A}$ is denoted $\mathrm{Der}\, \mathfrak{A}$.

It is clear that $\mathrm{Der}\, \mathfrak{A}$ is a subspace of $\mathrm{Hom}_k\, \mathfrak{A}$.

2.4.2 Proposition

Let $D \in \mathrm{Hom}_k\, \mathfrak{A}$. Then the following conditions are equivalent:

1. $D \in \mathrm{Der}\, \mathfrak{A}$;
2. $L_{xD} = [L_x, D]$ for $x \in \mathfrak{A}$;
3. $R_{xD} = [R_x, D]$ for $x \in \mathfrak{A}$.

PROOF. $(xy)D = (xD)y + x(yD) \Leftrightarrow yL_xD = yL_{xD} + yDL_x \Leftrightarrow yL_{xD} = y[L_x, D]$. Similarly, $(xy)D = (xD)y + x(yD) \Leftrightarrow xR_{yD} = x[R_y, D]$.

As a special case, suppose that $\mathfrak{A}$ is an associative algebra and $a \in \mathfrak{A}$. Recall that $x \operatorname{ad} a = [x, a] = xa - ax$ for $x \in \mathfrak{A}$, so that $\operatorname{ad} a = L_a - R_a$. Then $\operatorname{ad} a$ is a derivation of $\mathfrak{A}$, by 2.4.2, since $[L_x, \operatorname{ad} a] = [L_x, L_a - R_a] = [L_x, L_a] = L_{[x, a]} = L_{x \operatorname{ad} a}$. Also, $\operatorname{ad} a$ is a derivation of $\mathfrak{A}_{\text{Lie}}$, for $[x, y] \operatorname{ad} a = (xy - yx) \operatorname{ad} a = (x \operatorname{ad} a)y + x(y \operatorname{ad} a) - (y \operatorname{ad} a)x - y(x \operatorname{ad} a) = [x \operatorname{ad} a, y] + [x, y \operatorname{ad} a]$ for $x, y \in \mathfrak{A}$. Thus, we have:

2.4.3 Proposition
Let $\mathfrak{A}$ be an associative algebra. Then $\operatorname{ad} \mathfrak{A} \subset \operatorname{Der} \mathfrak{A}_{\text{Assoc}}$ and $\operatorname{ad} \mathfrak{A} \subset \operatorname{Der} \mathfrak{A}_{\text{Lie}}$.

2.4.4 Proposition
Let $\mathfrak{A}$ be anticommutative. Then the following conditions are equivalent:
1. $\mathfrak{A}$ is a Lie algebra;
2. $L_{\mathfrak{A}} \subset \operatorname{Der} \mathfrak{A}$;
3. $R_{\mathfrak{A}} \subset \operatorname{Der} \mathfrak{A}$.

PROOF. $(xy)z + (yz)x + (zx)y = 0 \Leftrightarrow x(yz) = (xy)z + y(xz) \Leftrightarrow (yz)L_x = (yL_x)z + y(zL_x)$. Thus, A is a Lie algebra $\Leftrightarrow L_{\mathfrak{A}} \subset \operatorname{Der} \mathfrak{A} \Leftrightarrow R_{\mathfrak{A}} = -L_{\mathfrak{A}} \subset \operatorname{Der} \mathfrak{A}$.

2.4.5 Corollary
Let $\mathfrak{A}$ be an associative algebra. Then $\mathfrak{A}_{\text{Lie}}$ is a Lie algebra.

PROOF. $\mathfrak{A}_{\text{Lie}}$ is anticommutative and we apply 2.4.3 and 2.4.4.

2.4.6 Proposition
$\operatorname{Der} \mathfrak{A}$ is a subalgebra of $(\operatorname{Hom}_k \mathfrak{A})_{\text{Lie}}$.

PROOF. Let D, E $\in \operatorname{Der} \mathfrak{A}$. Then
$[L_x, [D, E]] = [[L_x, D], E] + [D, [L_x, E]] = [L_{xD}, E] + [D, L_{xE}] = L_{xDE} - L_{xED} = L_{x[D, E]}$ for $x \in \mathfrak{A}$, and $[D, E] \in \operatorname{Der} \mathfrak{A}$ by 2.4.2.

2.4.7 Corollary
Der $\mathfrak{A}$ with product $[D, E] = DE - ED$ is a Lie algebra.

2.4.8 Proposition
Let $\mathfrak{A}$ be a Lie algebra. Then
1. $[\operatorname{ad} a, D] = \operatorname{ad} aD$ for $D \in \operatorname{Der} \mathfrak{A}$;
2. ad $\mathfrak{A}$ is an ideal of Der $\mathfrak{A}$;
3. $\operatorname{ad} : \mathfrak{A} \to \operatorname{Der} \mathfrak{A}$ is a homomorphism.

PROOF. 1 follows from 2.4.2, 2 follows from 1, and 3 from 1 and the identities $\operatorname{ad}(xy) = \operatorname{ad}(x \operatorname{ad} y) = [\operatorname{ad} x, \operatorname{ad} y]$.

2.4.9 Theorem
Let $D \in \operatorname{Der} \mathfrak{A}$ and $\alpha, \beta \in k$. Then

$$(xy)(D - \alpha - \beta)^n = \sum \binom{m}{n} x(D - \alpha)^m y(D - \beta)^{n-m}$$

for all positive integers n and $x, y \in \mathfrak{A}$.

PROOF. We prove this by induction on n. For $n = 1$, it is true since

$$(xy)(D - \alpha - \beta) = xDy - \alpha(xy) + x(yD) - \beta(xy).$$

Next, if

$$(xy)(D - \alpha - \beta)^{n-1} = \sum \binom{n-1}{m} x(D - \alpha)^m y(D - \beta)^{n-m-1},$$

then

$$\begin{aligned}(xy)(D - \alpha - \beta)^n &= \sum \binom{n-1}{m} x(D - \alpha)^{m+1} y(D - \beta)^{n-(m+1)} \\ &\quad + \sum \binom{n-1}{m} x(D - \alpha)^m y(D - \beta)^{n-m} \\ &= \sum \binom{n}{m} x(D - \alpha)^m y(D - \beta)^{n-m},\end{aligned}$$

since $\binom{n-1}{m} + \binom{n-1}{m+1} = \binom{n}{m+1}$ for $m < n$.

2.4.10 Corollary (Leibniz's rule)
For $D \in \operatorname{Der} \mathfrak{A}$,

$$(xy)D^n = \sum \binom{n}{m} xD^m yD^{n-m}$$

for all $x, y \in \mathfrak{A}$.

2.4.11 Corollary
If the characteristic of k is $p > 0$, then $D \in \operatorname{Der} \mathfrak{A}$ implies $D^p \in \operatorname{Der} \mathfrak{A}$.

PROOF. $\sum \binom{p}{m} xD^m yD^{p-m} = (xD^p)y + 0 + \ldots + 0 + x(yD^p)$.

2.4.12 Corollary
Let $\mathfrak{A}$ be finite dimensional, $D \in \operatorname{Der} \mathfrak{A}$. Then $\mathfrak{A}_\alpha(D)\mathfrak{A}_\beta(D) \subset \mathfrak{A}_{\alpha+\beta}(D)$ for $\alpha, \beta \in k$, $\mathfrak{A}_0(D)$ is a subalgebra of $\mathfrak{A}$, and $\mathfrak{A}_0(D)\mathfrak{A}_*(D) \subset \mathfrak{A}_*(D)$, $\mathfrak{A}_*(D)\mathfrak{A}_0(D) \subset \mathfrak{A}_*(D)$.

PROOF. Let $x \in \mathfrak{A}_\alpha(D)$, $y \in \mathfrak{A}_\beta(D)$ and choose r such that $x(D - \alpha)^r = y(D - \beta)^r = 0$. Letting $n = 2r$, $(xy)(D - \alpha - \beta)^n = 0$ by 2.4.9. Thus, $xy \in \mathfrak{A}_{\alpha+\beta}(D)$ and $\mathfrak{A}_\alpha(D)\mathfrak{A}_\beta(D) \subset \mathfrak{A}_{\alpha+\beta}(D)$. In particular, $\mathfrak{A}_0(D)$ is a subalgebra of $\mathfrak{A}$. Now let k' be the algebraic closure of k and let $\mathfrak{A}' = \mathfrak{A}_{k'}$. Then $\mathfrak{A}'_*(D) = \sum_{\alpha \neq 0} \mathfrak{A}'_\alpha(D)$, so $\mathfrak{A}'_0(D)\mathfrak{A}'_*(D) \subset \mathfrak{A}'_*(D)$ and $\mathfrak{A}'_*(D)\mathfrak{A}'_0(D) \subset \mathfrak{A}'_*(D)$ by the above observations. Since $\mathfrak{A}_0(D)$, $\mathfrak{A}_*(D)$ are k-forms of $\mathfrak{A}'_0(D)$, $\mathfrak{A}'_*(D)$, the same is true of them and all the assertions are proved.

2.4.13 Proposition
Let D be a split semisimple linear transformation of $\mathfrak{A}$, where $\mathfrak{A}$ is finite dimensional. Then $D \in \operatorname{Der} \mathfrak{A}$ iff $\mathfrak{A}_\alpha(D)\mathfrak{A}_\beta(D) \subset \mathfrak{A}_{\alpha+\beta}(D)$ for all $\alpha, \beta \in k$.

PROOF. One direction follows from 2.4.12. Next, assume that $\mathfrak{A}_\alpha(D)\mathfrak{A}_\beta(D) \subset \mathfrak{A}_{\alpha+\beta}(D)$ for $\alpha, \beta \in k$. Let $x \in \mathfrak{A}_\alpha(D)$, $y \in \mathfrak{A}_\beta(D)$. Then $xy \in \mathfrak{A}_{\alpha+\beta}(D)$ and we have $(xy)D = (\alpha + \beta)xy = (\alpha x)y + x(\beta y) = (xD)y + x(yD)$. Since $\bigcup_{\alpha \in k} \mathfrak{A}_\alpha(D)$ spans $\mathfrak{A}$, it follows that $D \in \operatorname{Der} \mathfrak{A}$.

2.4.14 Corollary

Let $D \in \operatorname{Der} \mathfrak{A}$, where $\mathfrak{A}$ is finite dimensional. Then if D is split or separable over k, D_s and D_n are contained in $\operatorname{Der} \mathfrak{A}$.

PROOF. Suppose first that D is split. Then $\mathfrak{A}_\alpha(D) = \mathfrak{A}_\alpha(D_s)$ for $\alpha \in k$, and $D_s \in \operatorname{Der} \mathfrak{A}$ by 2.4.13. Thus, $D_n = D - D_s \in \operatorname{Der} \mathfrak{A}$.

Next, suppose that D is separable over k, let k' be an algebraic Galois extension of k such that D is split over k', and let $\mathfrak{A}' = \mathfrak{A}_{k'}$. Then $D \in \operatorname{Der} \mathfrak{A}'$, so that D_s, $D_n \in \operatorname{Der} \mathfrak{A}'$ by the above paragraph. But D_s, D_n are defined over k by 1.4.10. Thus, D_s, $D_n \in \operatorname{Der} \mathfrak{A}$.

2.4.15 Proposition

Let D be a nilpotent derivation of $\mathfrak{A}$, where $\mathfrak{A}$ is finite dimensional over k and k is of characteristic 0. Then $\exp D = \sum_0^\infty (D^m/m!)$ is an automorphism of $\mathfrak{A}$.

PROOF. Choose n such that $D^{n+1} = 0$. Then

$$(xy) \exp D = \sum_{m=0}^{n} (xy)\frac{D^m}{m!} = \sum_{m=0}^{n} \sum_{i=0}^{m} \frac{1}{m!}\binom{m}{i} xD^i yD^{m-i}$$

by Leibniz's rule. But

$$x \exp D\, y \exp D = \sum_{j=0}^{n} \sum_{i=0}^{n} \frac{xD^i}{i!}\frac{yD^j}{j!}$$

$$= \sum_{m=0}^{2n} \sum_{i=0}^{m} \frac{xD^i}{i!}\frac{yD^{m-i}}{(m-i)!},$$

since $xD^i = 0$ for $i > n$. But we may replace n by $2n$ in the earlier equation, since $yD^i = 0$ for $i > n$, and it follows that $(xy) \exp D = x \exp D\, y \exp D$.

2.4.16 Definition

An ideal of $\mathfrak{A}$ is *characteristic* if it is stable under every derivation of $\mathfrak{A}$.

2.4.17 Theorem

Let $\mathfrak{B}$ be a minimal proper ideal of $\mathfrak{A}$. Then either $\mathfrak{B}$ is characteristic or $\mathfrak{B}^2 = 0$.

PROOF. Let $\mathfrak{C} = \{x \in \mathfrak{B} \mid x \operatorname{Der} \mathfrak{A} \subset \mathfrak{B}\}$. Then $\mathfrak{C}$ is an ideal of $\mathfrak{A}$ since $\mathfrak{B}$ is an ideal of $\mathfrak{A}$, for $x \in \mathfrak{B}$ with $xD \in \mathfrak{B}$ implies $(xy)D = (xD)y + x(yD) \in \mathfrak{B}$ and $(yx)D \in \mathfrak{B}$ for $D \in \operatorname{Der} \mathfrak{A}$ and $y \in \mathfrak{A}$. And $\mathfrak{B}^2 \subset \mathfrak{C} \subset \mathfrak{B}$, since $(xy)D = (xD)y + x(yD) \in \mathfrak{B}$ for $x, y \in \mathfrak{B}$.

Now $\mathfrak{C} = \{0\}$ or $\mathfrak{C} = \mathfrak{B}$, by the minimality of $\mathfrak{B}$. Thus, $\mathfrak{B}^2 = 0$ or $\mathfrak{C} = \mathfrak{B}$. In the latter case, $\mathfrak{B}$ is characteristic.

2.5 Solvability; the Radical of $\mathfrak{A}$

2.5.1 Definition

The subalgebras $\mathfrak{A}^{(i)}$, $i \geq 0$, are recursively defined by $\mathfrak{A}^{(0)} = \mathfrak{A}$, $\mathfrak{A}^{(i)} = \mathfrak{A}^{(i-1)}\mathfrak{A}^{(i-1)}$ for $i \geq 1$.

If $\mathfrak{A}$ is a Lie algebra, the $\mathfrak{A}^{(i)}$ are ideals, by 2.4.4.

2.5.2 Definition

The ideals $\mathfrak{A}^i$, $i \geq 1$, are recursively defined by $\mathfrak{A}^1 = \mathfrak{A}$, $\mathfrak{A}^i = \mathfrak{A}^{i-1}M(\mathfrak{A})$ for $i \geq 2$.

2.5.3 Definition

$\mathfrak{A}$ is *solvable* if $\mathfrak{A}^{(i)} = \{0\}$ for some i.

2.5.4 Definition

$\mathfrak{A}$ is *nilpotent* if $\mathfrak{A}^i = \{0\}$ for some i.

If $\mathfrak{A}$ is associative, $\mathfrak{A}$ is solvable if and only if $\mathfrak{A}$ is nilpotent, and the notion of nilpotency is equivalent to the usual notion of nilpotency.

2.5.5 Proposition

Let k' be an extension field of k. Then $\mathfrak{A}$ is solvable (respectively nilpotent) iff $\mathfrak{A}_{k'}$ is solvable (respectively nilpotent).

PROOF. Recall that $(\mathfrak{B}\mathfrak{C})_{k'} = \mathfrak{B}_{k'}\mathfrak{C}_{k'}$ for subspaces $\mathfrak{B}$, $\mathfrak{C}$ of $\mathfrak{A}$ by 2.2.3.

Thus, $(\mathfrak{A}^{(i)})_{k'} = (\mathfrak{A}_{k'})^{(i)}$, $i \geq 0$, and $\mathfrak{A}$ is solvable iff $\mathfrak{A}_{k'}$ is solvable. Similarly, $(\mathfrak{A}^{i})_{k'} = (\mathfrak{A}_{k'})^{i}$, $i \geq 1$, and $\mathfrak{A}$ is nilpotent iff $\mathfrak{A}_{k'}$ is nilpotent.

2.5.6 Proposition

Let $\mathfrak{B}$ be an ideal of $\mathfrak{A}$. Then $\mathfrak{A}$ is solvable iff $\mathfrak{B}$ and $\mathfrak{A}/\mathfrak{B}$ are solvable.

PROOF. It is clear that $(\mathfrak{A}/\mathfrak{B})^{(i)} = \mathfrak{A}^{(i)} + \mathfrak{B}/\mathfrak{B}$ and $\mathfrak{B}^{(i)} \subset \mathfrak{A}^{(i)}$. Thus, $\mathfrak{B}$ and $\mathfrak{A}/\mathfrak{B}$ are solvable if $\mathfrak{A}$ is solvable. If $\mathfrak{B}$ and $\mathfrak{A}/\mathfrak{B}$ are solvable, choose i, j such that $(\mathfrak{A}/\mathfrak{B})^{(i)} = \mathfrak{B}/\mathfrak{B}$ and $\mathfrak{B}^{(j)} = \{0\}$. Then $\mathfrak{A}^{(i)} \subset \mathfrak{B}$ and $\mathfrak{A}^{(i+j)} = (\mathfrak{A}^{(i)})^{(j)} = \{0\}$.

2.5.7 Corollary

Let $\mathfrak{B}$ be a solvable ideal of $\mathfrak{A}$, $\mathfrak{C}$ a solvable subalgebra of $\mathfrak{A}$. Then $\mathfrak{B} + \mathfrak{C}$ is solvable.

PROOF. $\mathfrak{B}$ is solvable, and $\mathfrak{B} + \mathfrak{C}/\mathfrak{B}$ is solvable since $\mathfrak{C}$ is solvable. Now apply 2.5.6.

2.5.8 Corollary

Let $\mathfrak{A}$ be finite dimensional. Then $\mathfrak{A}$ has a unique maximal solvable ideal.

PROOF. Let $\mathfrak{B}$, $\mathfrak{C}$ be maximal solvable ideals of $\mathfrak{A}$. Then $\mathfrak{B} + \mathfrak{C}$ is a solvable ideal of $\mathfrak{A}$, by 2.5.7, so that $\mathfrak{B} = \mathfrak{B} + \mathfrak{C} = \mathfrak{C}$.

2.5.9 Definition

The maximal solvable ideal of a finite-dimensional nonassociative algebra $\mathfrak{A}$ is called the *radical* of $\mathfrak{A}$ and is denoted Rad $\mathfrak{A}$.

If $\mathfrak{A}$ is a finite-dimensional associative algebra, the radical of $\mathfrak{A}$ is the maximal nilpotent ideal of $\mathfrak{A}$. Even in this case, however, Rad $\mathfrak{A}$ need not be a characteristic ideal of $\mathfrak{A}$.

2.5.10 Example

Let $\mathfrak{A}$ be an algebra over a field k of characteristic $p > 0$ with basis $1, x, \ldots, x^{p-1}$ such that $x^p = 1$. That is, $\mathfrak{A}$ is the group algebra of a cyclic group of order p. Then the radical of $\mathfrak{A}$ is Rad $\mathfrak{A} = \{\sum_{i=0}^{p-1} \gamma_i x^i \mid \Sigma\gamma_i =$

$0\}$, for this ideal is nilpotent because $y^p = 0$ for all its elements y, and is maximal nilpotent since it is of codimension one. It is not characteristic. For example, let D be defined by $D(\sum_0^{p-1} \gamma_i x^i) = \sum_0^{p-1} i\gamma_i x^{i-1}$. Then $D(x^i) = ix^{i-1}$ for all $i \geq 0$, since $x^p = 1$, and D is a derivation of $\mathfrak{A}$. Now Rad $\mathfrak{A}$ is not characteristic, since $x - 1 \in \text{Rad}\,\mathfrak{A}$ but $(x-1)D = xD = 1 \notin \text{Rad}\,\mathfrak{A}$.

We can prove that Rad $\mathfrak{A}$ is characteristic if $\mathfrak{A}$ is a finite-dimensional nonassociative algebra of characteristic 0. We first prove a cumbersome lemma.

2.5.11 Definition

The mappings f_m, $m \geq 0$, with domain the 2^m-fold Cartesian product of $\mathfrak{A}$ are, recursively, $f_0(x) = x$ for $x \in \mathfrak{A}$,

$$f_m(x_1, \ldots, x_{2^m}) = f_{m-1}(x_1, \ldots, x_{2^{m-1}}) f_{m-1}(x_{2^{m-1}+1}, \ldots, x_{2^m})$$

where $x_1, \ldots, x_{2^m} \in \mathfrak{A}$.

2.5.12 Lemma

Let $\mathfrak{B}$ be an ideal of $\mathfrak{A}$, and let $D \in \text{Der}\,\mathfrak{A}$. Then $f_n(x_1, \ldots, x_{2^n})D^{2^n} \equiv 2^n! f_n(x_1 D, \ldots, x_{2^n} D) (\text{mod}\ \mathfrak{B})$ and $f_n(x_1, \ldots, x_{2^n})D^m \in \mathfrak{B}$ for $m < 2^n$ and $x_i \in \mathfrak{B}$.

PROOF. The proof is by induction on n and is trivial if $n = 0$ or $n = 1$. For $n > 1$ and $m \leq 2^n$,

$$\sum_{r+s=m} \binom{m}{r} f_{n-1}(x_1, \ldots, x_{2^{n-1}}) D^r f_{n-1}(x_{2^{n-1}+1}, \ldots, x_{2^n}) D^s$$
$$= f_n(x_1, \ldots, x_{2^n}) D^m,$$

by Leibniz's rule. If $m < 2^n$, either $r < 2^{n-1}$ or $s < 2^{n-1}$ for $r + s = m$, and the expression is in $\mathfrak{B}$ by induction. Next, let $m = 2^n$. By a similar argument, the only term of the sum which is possibly not in $\mathfrak{B}$ is the term

$$\binom{2^n}{2^{n-1}} f_{n-1}(x_1, \ldots, x_{2^{n-1}})D^{2^{n-1}} f_{n-1}(x_{2^{n-1}+1}, \ldots, x_{2^n})D^{2^{n-1}}.$$

But this, by induction, is congruent modulo $\mathfrak{B}$ to

$$(2^{n-1})!(2^{n-1})!\binom{2^n}{2^{n-1}} f_{n-1}(x_1 D, \ldots, x_{2^{n-1}}D)$$

$$\times f_{n-1}(x_{2^{n-1}+1}D, \ldots, x_{2^n}D) = 2^n! f_n(x_1 D, \ldots, x_{2^n}D).$$

2.5.13 Theorem

Let $\mathfrak{A}$ be finite dimensional of characteristic 0. Then Rad $\mathfrak{A}$ is characteristic.

PROOF. Let $\mathfrak{R} =$ Rad $\mathfrak{A}$ and choose n such that $\mathfrak{R}^{(n)} = \{0\}$. We claim that the ideal $\mathfrak{R} + \mathfrak{R}D$ is solvable for $D \in$ Der $\mathfrak{A}$. For this, it suffices to show that $(\mathfrak{R} + \mathfrak{R}D)^{(n)} \subset \mathfrak{R}$, for then $(\mathfrak{R} + \mathfrak{R}D)^{(2n)} = \{0\}$. Now any element of $(\mathfrak{R} + \mathfrak{R}D)^{(n)}$ is a sum of elements of $\mathfrak{R}$ and elements of the form $f_n(x_1 D, \ldots, x_{2^n}D)(x_1, \ldots, x_{2^n} \in \mathfrak{R})$. But the latter elements are in $\mathfrak{R}$, by 2.5.12, for

$$f_n(x_1 D, \ldots, x_{2^n}D) \equiv \frac{1}{2^n!} f_n(x_1, \ldots, x_{2^n})D^{2^n} \equiv \frac{1}{2^n!}(0D^{2^n}) \equiv 0 \pmod{\mathfrak{R}}.$$

2.6 Simplicity and Semisimplicity

2.6.1 Definition

$\mathfrak{A}$ is *simple* if $\mathfrak{A}^2 = \mathfrak{A}$ and $\mathfrak{A}$ has no proper ideal.

2.6.2 Definition

$\mathfrak{A}$ is *characteristically simple* if $\mathfrak{A}^2 = \mathfrak{A}$ and $\mathfrak{A}$ has no proper characteristic ideal.

Since $\mathfrak{A}^2$ is a characteristic ideal of $\mathfrak{A}$, the condition $\mathfrak{A}^2 = \mathfrak{A}$ in the above definitions may be replaced by the condition $\mathfrak{A}^2 \neq \{0\}$, except of course when $\mathfrak{A} = \{0\}$.

2.6.3 Definition

A *simple ideal* of $\mathfrak{A}$ is an ideal $\mathfrak{B}$ of $\mathfrak{A}$ which is simple as a nonassociative algebra.

2.6.4 Proposition

$\mathfrak{A}$ is simple iff $\mathfrak{A}$ is $M(\mathfrak{A})$-irreducible. And $\mathfrak{A}$ is the direct sum of simple ideals of $\mathfrak{A}$ iff $\mathfrak{A}$ is $M(\mathfrak{A})$–completely reducible.

PROOF. The first assertion is obvious, as is one direction of the second. Next, let $\mathfrak{A}$ be $M(\mathfrak{A})$–completely reducible. Then $\mathfrak{A} = \sum \oplus \mathfrak{A}_i$ where the $\mathfrak{A}_i$ are $M(\mathfrak{A})$-irreducible $M(\mathfrak{A})$-submodules of $\mathfrak{A}$. Thus, the $\mathfrak{A}_i$ are ideals of $\mathfrak{A}$. We claim that they are simple. Thus, note that $\mathfrak{A}_i\mathfrak{A}_j = \{0\}$ for $i \neq j$, since $\mathfrak{A}_i\mathfrak{A}_j \subset \mathfrak{A}_i \cap \mathfrak{A}_j$. Thus, an ideal of $\mathfrak{A}_i$ is an ideal of $\mathfrak{A}$, and $\mathfrak{A}_i^2 = \mathfrak{A}_i\mathfrak{A}$. It follows that $\mathfrak{A}_i$ has no proper ideal and $\mathfrak{A}_i^2 = \mathfrak{A}_i$ for all i.

2.6.5 Corollary

Let $\mathfrak{A}$ be finite dimensional and $\mathfrak{A} = \sum \oplus \mathfrak{A}_i$ where the $\mathfrak{A}_i$ are simple ideals of $\mathfrak{A}$. Then every ideal $\mathfrak{B}$ of $\mathfrak{A}$ is a sum of certain of the $\mathfrak{A}_i$. In particular, the $\mathfrak{A}_i$ are the only simple ideals of $\mathfrak{A}$.

PROOF. $\mathfrak{A}$ is $M(\mathfrak{A})$–completely reducible, by 2.6.4. Thus, $\mathfrak{B}$ has an $M(\mathfrak{A})$-complement $\mathfrak{C}$. Now $\mathfrak{C}$ is an ideal of $\mathfrak{A}$ and $\mathfrak{A} = \mathfrak{B} \oplus \mathfrak{C}$. Since $\mathfrak{A} = \mathfrak{A}^2$, we have $\mathfrak{A} = \mathfrak{B}\mathfrak{A} \oplus \mathfrak{C}\mathfrak{A}$ and $\mathfrak{B} = \mathfrak{B}\mathfrak{A} = \sum \mathfrak{B}\mathfrak{A}_i$. But $\mathfrak{B}\mathfrak{A}_i$ is an ideal of $\mathfrak{A}_i$ so that $\mathfrak{B}\mathfrak{A}_i = \{0\}$ or $\mathfrak{B}\mathfrak{A}_i = \mathfrak{A}_i$. Thus, $\mathfrak{B}$ is the sum of those $\mathfrak{A}_i$ with $\mathfrak{B}\mathfrak{A}_i = \mathfrak{A}_i$.

2.6.6. Definition

$\mathfrak{A}$ is *semisimple* if $\mathfrak{A}$ has no nonzero solvable ideal.

If $\mathfrak{A}$ is finite dimensional, then $\mathfrak{A}/\text{Rad}\,\mathfrak{A}$ is semisimple, and $\mathfrak{A}$ is semisimple iff $\text{Rad}\,\mathfrak{A} = \{0\}$. If $\mathfrak{A}$ is a direct sum of simple ideals, then $\mathfrak{A}$ is semisimple by 2.6.5. The converse of this is not always true, but is true provided that $\mathfrak{A}$ possess a suitable nondegenerate bilinear form. We discuss this further in 2.7.

The characteristically simple nonassociative algebras are closely related to the simple nonassociative algebras. A consequence of a recent result of

Richard Block [4] is that a finite-dimensional characteristically simple nonassociative algebra $\mathfrak{A}$ over k has the form $\mathfrak{A} = \mathfrak{B} \otimes_k k[G]$ where $\mathfrak{B}$ is a simple nonassociative algebra over k and $k[G]$ is the group algebra of a finite Abelian group of exponent p. The situation is particularly clear in the following two cases.

2.6.7 Theorem

Let $\mathfrak{A}$ be finite dimensional and semisimple. Then $\mathfrak{A}$ is characteristically simple iff $\mathfrak{A}$ is simple.

PROOF. Let $\mathfrak{A}$ be characteristically simple, $\mathfrak{B}$ a minimal nonzero ideal of $\mathfrak{A}$. Then $\mathfrak{B}^2 \neq \{0\}$ since $\mathfrak{A}$ is semisimple. Thus, $\mathfrak{B}$ is characteristic by 2.4.17. Thus, $\mathfrak{B} = \mathfrak{A}$ and $\mathfrak{A}$ is simple.

2.6.8 Theorem

Let $\mathfrak{A}$ be finite dimensional of characteristic 0. Then $\mathfrak{A}$ is characteristically simple iff $\mathfrak{A}$ is simple.

PROOF. Suppose that $\mathfrak{A}$ is nonzero and characteristically simple. Since the radical $\mathfrak{R}$ of $\mathfrak{A}$ is characteristic by 2.5.13, $\mathfrak{R} = \mathfrak{A}$ or $\mathfrak{R} = \{0\}$. Thus, $\mathfrak{A}$ is either solvable or semisimple. If $\mathfrak{A}$ is solvable, then $\mathfrak{A} = \{0\}$ since $\mathfrak{A}^2$ is a nonzero characteristic ideal. If $\mathfrak{A}$ is semisimple, then $\mathfrak{A}$ is simple by 2.6.7.

2.7 Nonassociative Algebras with Invariant Forms

We assume now that $\mathfrak{A}$ is finite dimensional.

2.7.1 Definition

An *associative form* on $\mathfrak{A}$ is a bilinear form $(\ ,\ )$ on $\mathfrak{A}$ such that $(xy, z) = (x, yz)$ for $x, y, z \in \mathfrak{A}$. An *invariant form* on $\mathfrak{A}$ is a symmetric associative form on $\mathfrak{A}$.

If $(\ ,\ )$ is a symmetric bilinear form on a vector space V over k, we say that $v \perp w$ if $(v, w) = 0$ and $v \perp S$ if $v \perp w$ for $w \in S$. Here we take $v \in V$, $w \in V$, $S \subset V$. We let $S^\perp = \{v \in V \mid v \perp S\}$. The *radical* of V is $V^\perp$. If $V^\perp = \{0\}$, $(\ ,\ )$ is *nondegenerate*. We can pass from V with $(\ ,\ )$ to $V' =$

$V/V^{\perp}$ with $(\ ,\ )'$, where $(v + V^{\perp}, w + V^{\perp})' = (v, w)$ for $v, w \in V$. Then $(\ ,\ )'$ is nondegenerate on V'.

2.7.2 Proposition

Let $(\ ,\ )$ be an invariant form on $\mathfrak{A}$, $\mathfrak{B}$ an ideal of $\mathfrak{A}$. Then $\mathfrak{B}^{\perp}$ is an ideal of $\mathfrak{A}$.

PROOF. Let $x \in \mathfrak{B}^{\perp}$, $y \in \mathfrak{A}$. Then $(xy, z) = (x, yz) = 0$ for $z \in B$ and $xy \perp \mathfrak{B}$. Similarly, $yx \perp \mathfrak{B}$ and $\mathfrak{B}^{\perp}$ is an ideal of $\mathfrak{A}$.

2.7.3 Definition

$\mathfrak{A}$ is a *symmetric algebra* if $\mathfrak{A}$ has a nondegenerate invariant form.

If $(\ ,\ )$ is an invariant form for $\mathfrak{A}$, then $\mathfrak{A}/\mathfrak{A}^{\perp}$ is a symmetric algebra.

2.7.4 Theorem (Dieudonné)

$\mathfrak{A}$ is a semisimple symmetric algebra iff $\mathfrak{A} = \sum \oplus \mathfrak{A}_i$ where the $\mathfrak{A}_i$ are simple ideals that are symmetric algebras. If $\mathfrak{A}$ is semisimple and $(\ ,\ )$ is a nondegenerate invariant form on $\mathfrak{A}$, then $\mathfrak{A}_i \perp \mathfrak{A}_j$ for $i \neq j$, where the $\mathfrak{A}_i$ are the simple ideals of $\mathfrak{A}$.

PROOF. If $\mathfrak{A} = \sum \oplus \mathfrak{A}_i$ where the $\mathfrak{A}_i$ are simple ideals which are symmetric algebras, then $\mathfrak{A}$ is semisimple by 2.6.6, and we obtain a nondegenerate invariant form for $\mathfrak{A}$ by taking the sum of nondegenerate invariant forms for the $\mathfrak{A}_i$.

Suppose conversely that $\mathfrak{A}$ is semisimple with nondegenerate invariant form $(\ ,\ )$. Let $\mathfrak{B}$ be an ideal of $\mathfrak{A}$. We claim that $\mathfrak{A} = \mathfrak{B} \oplus \mathfrak{B}^{\perp}$. Since $\dim \mathfrak{B} + \dim \mathfrak{B}^{\perp} = \dim \mathfrak{A}$, it suffices to show that $\mathfrak{C} = \mathfrak{B} \cap \mathfrak{B}^{\perp}$ is $\{0\}$. Since $\mathfrak{C}$ is an ideal, it therefore suffices to show that $\mathfrak{C}^2 = \{0\}$, since $\mathfrak{A}$ is semisimple. Now $\mathfrak{C} \subset \mathfrak{B}$ and $\mathfrak{C}\mathfrak{A} \subset \mathfrak{B}^{\perp}$, so $(cc', a) = (c, c'a) = 0$ for $c, c' \in \mathfrak{C}$ and $a \in \mathfrak{A}$. Thus, $\mathfrak{C}^2 \subset \mathfrak{A}^{\perp} = \{0\}$ and $\mathfrak{C}^2 = \{0\}$. Thus, $\mathfrak{C} = \{0\}$ and $\mathfrak{A} = \mathfrak{B} \oplus \mathfrak{B}^{\perp}$. It follows that $\mathfrak{A}$ is $M(\mathfrak{A})$–completely reducible, in fact that $\mathfrak{A} = \mathfrak{A}_1 \oplus \ldots \oplus \mathfrak{A}_n$ where the $\mathfrak{A}_i$ are orthogonal minimal ideals such that $\mathfrak{A}_i\mathfrak{A}_j = \{0\}$ for $i \neq j$. The $\mathfrak{A}_i$ are the simple ideals of $\mathfrak{A}$, as in 2.6.5, and all of the assertions follow.

We will use the following theorem to prove an extremely important result of Zassenhaus, 2.8.6.

2.7.5 Theorem

Let $(,)$ be an invariant form on $\mathfrak{A}$ and let $\mathfrak{B}$ be an ideal of $\mathfrak{A}$ such that $(,)|_{\mathfrak{B}}$ is nondegenerate. Then for $a \in \mathfrak{A}$, there exists $b \in \mathfrak{B}$ such that $L_a|_{\mathfrak{B}} = L_b|_{\mathfrak{B}}$ and $R_a|_{\mathfrak{B}} = R_b|_{\mathfrak{B}}$.

PROOF. The hypothesis implies that $\mathfrak{A} = \mathfrak{B} \oplus \mathfrak{B}^{\perp}$. Now for $a \in \mathfrak{A}$, let b be the corresponding projection of a on $\mathfrak{B}$. Thus, $a - b \in \mathfrak{B}^{\perp}$. Since $\mathfrak{B}$ and $\mathfrak{B}^{\perp}$ are ideals of $\mathfrak{A}$, $\mathfrak{B}\mathfrak{B}^{\perp} = \mathfrak{B}^{\perp}\mathfrak{B} = \{0\}$. Thus, $L_a|_{\mathfrak{B}} = L_b|_{\mathfrak{B}}$ and $R_a|_{\mathfrak{B}} = R_b|_{\mathfrak{B}}$.

2.8 Lie Algebras with Invariant Forms

We now let $\mathfrak{L}$ be a finite-dimensional Lie algebra over k, V a finite-dimensional vector space over k, and $f\colon \mathfrak{L} \to (\mathrm{Hom}_k\, V)_{\mathrm{Lie}}$ a Lie algebra homomorphism. We introduce here an important class of invariant forms for $\mathfrak{L}$ obtained by taking such V and f.

2.8.1 Definition

The form $\mathrm{T}_f(x, y) = \mathrm{Trace}\, f(x)f(y)$ on $\mathfrak{L}$ is called the *trace form* of $\mathfrak{L}$ with respect to f.

2.8.2 Proposition

$\mathrm{T}_f(,)$ is an invariant form on $\mathfrak{L}$.

PROOF.

$\mathrm{T}_f(xy, z) = \mathrm{Trace}\, f(xy)f(z)$
$= \mathrm{Trace}\, [f(x), f(y)]f(z) = \mathrm{Trace}\, f(x)f(y)f(z)$
$- \mathrm{Trace}\, f(y)f(x)f(z) = \mathrm{Trace}\, f(x)f(y)f(z)$
$- \mathrm{Trace}\, (f(x)f(z))f(y) = \mathrm{Trace}\, f(x)[f(y), f(z)]$
$= \mathrm{Trace}\, f(x)f(yz) = \mathrm{T}_f(x, yz)$.

2.8.3 Definition

$\mathrm{K}(,) = \mathrm{T}_{\mathrm{ad}}(,)$.

By 2.8.2, $\mathrm{K}(,)$ is an invariant form. It is called the *Killing form* of $\mathfrak{L}$ and is given by $\mathrm{K}(x, y) = \mathrm{Trace}\ \mathrm{ad}\, x\ \mathrm{ad}\, y$.

2.8.4 Proposition
If K (,) is nondegenerate on $\mathfrak{L}$, then $\mathfrak{L}$ is semisimple.

PROOF. Let $\mathfrak{B}$ be an ideal of $\mathfrak{L}$ with $\mathfrak{B}^2 = \{0\}$. Then $\mathfrak{B}$ ad $\mathfrak{B}$ ad $\mathfrak{L} = \{0\}$ and $\mathfrak{L}$ ad $\mathfrak{B}$ ad $\mathfrak{L} \subset \mathfrak{B}$. It follows that Trace ad b ad $x = 0$ for $b \in \mathfrak{B}$, $x \in \mathfrak{L}$, and $\mathfrak{B} \subset \mathfrak{L}^{\perp}$. Thus, $\mathfrak{B} = \{0\}$ and $\mathfrak{L}$ is semisimple, for if Rad $\mathfrak{L} \neq \{0\}$ we could take $\mathfrak{B} = (\text{Rad } \mathfrak{L})^{(i)} \neq \{0\}$, where $(\text{Rad } \mathfrak{L})^{(i+1)} = \{0\}$ for some i. But then $\mathfrak{B} \neq \{0\}$ and $\mathfrak{B}^2 = \{0\}$, a contradiction.

2.8.5 Corollary
Let the Killing form K (,) of $\mathfrak{L}$ be nondegenerate. Then $\mathfrak{L} = \sum \oplus \mathfrak{L}_i$ where the $\mathfrak{L}_i$ are simple ideals each of which have nondegenerate Killing form.

PROOF. Since $\mathfrak{L}_i \perp \mathfrak{L}_j$ for $i \neq j$, by 2.6.5, the restriction of K (,) to $\mathfrak{L}_i$ is nondegenerate and is the Killing form of $\mathfrak{L}_i$.

2.8.6 Theorem (Zassenhaus)
Let $\mathfrak{L}$ have nondegenerate Killing form. Then ad $\mathfrak{L}$ = Der $\mathfrak{L}$.

PROOF. Let $(D, E) =$ Trace DE for $D, E \in$ Der $\mathfrak{L}$. Then (,) is an invariant form on Der $\mathfrak{L}$, and $(\ )|_{\text{ad } \mathfrak{L}}$ is nondegenerate since it is the Killing form. Thus, by 2.7.5, for each $D \in$ Der $\mathfrak{L}$ there exists ad $x \in$ ad $\mathfrak{L}$ such that $[\text{ad } y, D] = [\text{ad } y, \text{ad } x]$ for all ad $y \in$ ad $\mathfrak{L}$. Thus $\text{ad}(yD) = \text{ad}(y \text{ ad } x)$ for all $y \in \mathfrak{L}$. To show that $D = \text{ad } x$, it therefore suffices to show that ad is injective or that $\mathfrak{B} =$ Kernel ad is $\{0\}$. But $\mathfrak{B}$ is an ideal such that $\mathfrak{B}\mathfrak{L} = \{0\}$, so $\mathfrak{B}^2 = \{0\}$. Since $\mathfrak{L}$ is semisimple, $\mathfrak{B} = \{0\}$.

3

Lie Algebras of Characteristic 0

3.1 Introduction

Finite-dimensional Lie algebras of characteristic 0 are the most widely and successfully studied Lie algebras, mainly because of the close connection that they have with algebraic groups and Lie groups. This relationship breaks down at characteristic $p > 0$ since the exponential mapping $x \mapsto \sum(x^n/n!)$ involves coefficients $1/p$. For $p > 0$, the theory of Lie algebras and theory of algebraic groups diverge, the theory of Lie algebras becoming extremely complicated and the theory of algebraic groups remaining, in comparison, much the same.

In this chapter, we briefly cover the classical theorems on Lie algebras of characteristic 0. These theorems give the general structure of Lie algebras of characteristic 0 and give the classification of split semisimple Lie algebras of characteristic 0. The development of the material is designed to emphasize the simplicity of the subject rather than to set up the groundwork for a highly comprehensive account. For more on Lie algebras of characteristic 0, we refer the reader to [13].

The contents of this chapter are roughly as follows. Nilpotency criteria are given in terms of right translations (Engel's theorem). Criteria for solvability and semisimplicity are given in terms of the Killing form (Cartan's criteria). A decomposition $\mathfrak{L} = \mathfrak{B} \oplus \operatorname{Rad} \mathfrak{L}$ is given which shows how a Lie algebra $\mathfrak{L}$ is built up from a semisimple subalgebra $\mathfrak{B}$ and the solvable ideal $\operatorname{Rad} \mathfrak{L}$ (Levi's theorem). Some basic results on representations of nilpotent, solvable, and semisimple Lie algebras are given. Finally, the theory of abstract root systems is developed and used to give the classification of split semisimple Lie algebras of characteristic 0 and a description of their automorphism groups.

We begin with the basic language and some preliminary observations. Throughout the chapter except where otherwise indicated, k is a field of characteristic 0 and $\mathfrak{L}$ is a finite-dimensional Lie algebra over k. A few results are formulated for Lie algebras of arbitrary characteristic for use in Chapter 4.

3.1.1 Definition

If $\mathfrak{L}^2 = \{0\}$, we say that $\mathfrak{L}$ is *Abelian.* Two elements x, y of $\mathfrak{L}$ *commute* if $xy = 0$. For subsets $\mathfrak{S}$, $\mathfrak{B}$ of $\mathfrak{L}$, the *centralizer* of $\mathfrak{S}$ in $\mathfrak{B}$ is the set $\mathfrak{C}_{\mathfrak{B}}(\mathfrak{S})$

of elements of $\mathfrak{B}$ which commute with every element of $\mathfrak{S}$. The *centralizer* of $\mathfrak{S}$ is $\mathfrak{C}_{\mathfrak{L}}(\mathfrak{S})$. The *center* of $\mathfrak{L}$ is $\mathfrak{C}(\mathfrak{L}) = \mathfrak{C}_{\mathfrak{L}}(\mathfrak{L})$.

3.1.2 Definition

An element x of $\mathfrak{L}$ *normalizes* the subspace $\mathfrak{B}$ of $\mathfrak{L}$ if $x\mathfrak{B} \subset \mathfrak{B}$. The set of elements normalizing $\mathfrak{B}$ is denoted $\mathfrak{N}_{\mathfrak{L}}(\mathfrak{B})$ and is a subalgebra of $\mathfrak{L}$ called the *normalizer* of $\mathfrak{B}$.

Since $\mathfrak{L}$ is anticommutative, a subalgebra $\mathfrak{B}$ of $\mathfrak{L}$ is an ideal of $\mathfrak{N}_{\mathfrak{L}}(\mathfrak{B})$.

If $\mathfrak{B}$ and $\mathfrak{C}$ are ideals of $\mathfrak{L}$, then $\mathfrak{B}\mathfrak{C} = \mathfrak{C}\mathfrak{B}$ is an ideal of $\mathfrak{L}$, since right and left translations by elements of $\mathfrak{L}$ are derivations of $\mathfrak{L}$. In particular, the subalgebras $\mathfrak{L}^{(i)}$, $\mathfrak{L}^{i}$ introduced in 2.5 are ideals of $\mathfrak{L}$. The series $\mathfrak{L}^{(i)}$ is the *commutator series*, the series $\mathfrak{L}^{i}$ the *descending central series*. We recall that $\mathfrak{L}^{(i)}$ was defined by $\mathfrak{L}^{(0)} = \mathfrak{L}$, $\mathfrak{L}^{(i)} = \mathfrak{L}^{(i-1)}\mathfrak{L}^{(i-1)}$ for $i \geq 1$. The earlier definition of $\mathfrak{L}^{i}$ simplifies for Lie algebras to $\mathfrak{L}^{1} = \mathfrak{L}$, $\mathfrak{L}^{i} = \mathfrak{L}^{i-1}\mathfrak{L}$ for $i \geq 2$. We recall also that $\mathfrak{L}$ is *solvable* if $\mathfrak{L}^{(i)} = \{0\}$ for some i, *nilpotent* if $\mathfrak{L}^{i} = \{0\}$ for some i. One easily sees that $\mathfrak{L}^{(i)}/\mathfrak{L}^{(i+1)}$ is Abelian and that $\mathfrak{L}^{i-1}/\mathfrak{L}^{i}$ is central in $\mathfrak{L}/\mathfrak{L}^{i}$ for all i. As in the theory of groups, $\mathfrak{L}$ is solvable iff there exists a series $\mathfrak{L} = \mathfrak{L}_0 \supset \mathfrak{L}_1 \supset \ldots \supset \mathfrak{L}_n = \{0\}$ of ideals such that $\mathfrak{L}_i/\mathfrak{L}_{i+1}$ is Abelian for all i; similarly, $\mathfrak{L}$ is nilpotent iff there exists a series $\mathfrak{L} = \mathfrak{L}_1 \supset \mathfrak{L}_2 \supset \ldots \supset \mathfrak{L}_n = \{0\}$ such that $\mathfrak{L}_{i-1}/\mathfrak{L}_i$ is central in $\mathfrak{L}/\mathfrak{L}_i$ for all i.

3.1.3 Definition

A *representation* of a Lie algebra $\mathfrak{L}$ is a homomorphism f from $\mathfrak{L}$ into $(\mathrm{Hom}_k V)_{\mathrm{Lie}}$ for some finite-dimensional vector space V over k.

3.1.4 Definition

A *Lie module* for $\mathfrak{L}$ is an $\mathfrak{L}$-module $\mathfrak{M}$ over k such that $m(xy) = (mx)y - (my)x$ for $x, y \in \mathfrak{L}$, $m \in \mathfrak{M}$.

If $f: \mathfrak{L} \to (\mathrm{Hom}_k V)_{\mathrm{Lie}}$ is a representation of $\mathfrak{L}$, then V together with the module operation $mx = mf(x)$ for $m \in V$, $x \in \mathfrak{L}$ is a Lie module for $\mathfrak{L}$. Thus, $v(xy) = vf(xy) = v[f(x), f(y)] = (vf(x))f(y) - (vf(y))f(x) = (vx)y - (vy)x$ for $v \in V$, $x, y \in \mathfrak{L}$. The Lie module for $\mathfrak{L}$ obtained in this way is the *Lie module afforded by f*.

If $\mathfrak{M}$ is a Lie module for $\mathfrak{L}$ over k, then let $f: \mathfrak{L} \to \mathrm{Hom}_k\, \mathfrak{M}$ be defined by $vf(x) = vx$ for $x \in \mathfrak{L}$. Then f is a representation of $\mathfrak{L}$ and is called the *representation afforded by* $\mathfrak{M}$.

The representation $\mathrm{ad} : \mathfrak{L} \to \mathrm{Hom}_k\, \mathfrak{L}$ is the *adjoint representation* of $\mathfrak{L}$. The kernel of ad is $\mathfrak{C}(\mathfrak{L})$.

The direct sum of Lie modules for $\mathfrak{L}$ is a Lie module for $\mathfrak{L}$. If $\mathfrak{M}$ and $\mathfrak{N}$ are Lie modules for $\mathfrak{L}$, the tensor product $\mathfrak{M} \otimes_k \mathfrak{N}$ is a Lie module for $\mathfrak{L}$ when one introduces the module action such that $(m \otimes n)x = mx \otimes n + m \otimes nx$ for $m \in \mathfrak{M}$, $n \in \mathfrak{N}$, $x \in \mathfrak{L}$. (The tensor product of two Lie algebras is, however, not always a Lie algebra.)

Now let $\mathfrak{M}$ be a Lie module for $\mathfrak{L}$ and suppose that $\mathfrak{M}$ is also a Lie algebra such that $(mm')x = (mx)m' + m(m'x)$ for $m, m' \in \mathfrak{M}$ and $x \in \mathfrak{L}$. We then give $\mathfrak{L} \oplus \mathfrak{M}$ the nonassociative algebra product $(x \oplus m)(x' \oplus m') = xx' \oplus (mx' - m'x + mm')$ for $x, x' \in \mathfrak{L}$ and $m, m' \in \mathfrak{M}$. In particular, we give $\mathrm{Der}\, \mathfrak{L} \oplus \mathfrak{L}$ the product $(D \oplus x)(D' \oplus x') = [D, D'] \oplus xD' - x'D + xx'$ for $D, D' \in \mathrm{Der}\, \mathfrak{L}$ and $x, x' \in \mathfrak{L}$. The algebra $\mathfrak{L} \oplus \mathfrak{M}$, and therefore the algebra $\mathrm{Der}\, \mathfrak{L} \oplus \mathfrak{L}$, is a Lie algebra. For clearly $\mathfrak{L} \oplus \mathfrak{M}$ is anticommutative. And for $x \in \mathfrak{L}$, R_x is a derivation of $\mathfrak{L} \oplus \mathfrak{M}$ since $R_x|_{\mathfrak{L}}$ and $R_x|_{\mathfrak{M}}$ are derivations and $(my)x = (mx)y + m(yx)$ for $m \in \mathfrak{M}$, $y \in \mathfrak{L}$. Similarly, R_m is a derivation of $\mathfrak{L} \oplus \mathfrak{M}$ for $m \in \mathfrak{M}$, since $R_m|_{\mathfrak{M}}$ is a derivation, $(xy)m = -m(xy) = -(mx)y + (my)x = (xm)y + x(ym)$ for $x, y \in \mathfrak{L}$ and $(ny)m = (nm)y - n(my) = (nm)y + n(ym)$ for $y \in \mathfrak{L}$, $n \in \mathfrak{M}$. Thus, $R_{\mathfrak{L} \oplus \mathfrak{M}} \subset \mathrm{Der}\, \mathfrak{L} \oplus \mathfrak{M}$ and $\mathfrak{L} \oplus \mathfrak{M}$ is a Lie algebra by 2.4.4.

The Lie algebra $(\mathrm{Der}\, \mathfrak{L}) \oplus \mathfrak{L}$ is the *holomorph* of $\mathfrak{L}$, and $\mathfrak{M} \oplus \mathfrak{L}$ the *split extension of* $\mathfrak{L}$ *by* $\mathfrak{M}$.

3.2 Nilpotent Lie Algebras; Engel's Theorem

In this section, the characteristic of $\mathfrak{L}$ and k is arbitrary. Nilpotent Lie algebras are easily studied inductively by reducing questions to Abelian Lie algebras. Their role in the theory of Lie algebras is important because much of the structure of a Lie algebra can be described in terms of certain nilpotent subalgebras (Cartan subalgebras). We develop here some fundamental properties of nilpotent Lie algebras needed for this.

From the definition of nilpotency, the homomorphic image of a nilpotent Lie algebra is nilpotent. If $\mathfrak{C}$ is the center of $\mathfrak{L}$ and $\mathfrak{L}/\mathfrak{C}$ is nilpotent, then $\mathfrak{L}$ is nilpotent since $\mathfrak{L}^i \subset \mathfrak{C}$ implies $\mathfrak{L}^{i+1} = \{0\}$. Since $\mathfrak{C}$ is the kernel of ad, these observations give rise to the following proposition.

3.2.1 Proposition

$\mathfrak{L}$ is nilpotent iff ad $\mathfrak{L}$ is nilpotent.

3.2.2 Definition

The *ascending central series* of $\mathfrak{L}$ is the sequence $\mathfrak{C}^i(\mathfrak{L})$, $i \geq 0$, defined recursively by

$$\mathfrak{C}^0(\mathfrak{L}) = \{0\},\ \mathfrak{C}^i(\mathfrak{L}) = \{x \in \mathfrak{L} \mid x\mathfrak{L} \subset \mathfrak{C}^{i-1}(\mathfrak{L})\} \text{ for } i \geq 1.$$

Obviously $\mathfrak{C}^1(\mathfrak{L}) = \mathfrak{C}(\mathfrak{L})$ and $\mathfrak{C}^i(\mathfrak{L})/\mathfrak{C}^{i-1}(\mathfrak{L}) = \mathfrak{C}(\mathfrak{L}/\mathfrak{C}^{i-1}(\mathfrak{L}))$ for $i \geq 1$.

3.2.3 Proposition

$\mathfrak{L}^{i+1} = \{0\}$ iff $\mathfrak{C}^i(\mathfrak{L}) = \mathfrak{L}$ for $i \geq 0$.

PROOF. The details are straightforward and are left to the reader.

We now describe much more far-reaching nilpotency criteria.

3.2.4 Definition

Let $\mathfrak{N}$ be a subalgebra of $\mathfrak{L}$. Then $\mathfrak{C}^i_{\mathfrak{L}}(\mathfrak{N})$ is the series of subspaces of $\mathfrak{L}$ defined recursively by

$$\mathfrak{C}^0_{\mathfrak{L}}(\mathfrak{N}) = \{0\},\ \mathfrak{C}^i_{\mathfrak{L}}(\mathfrak{N}) = \{x \in \mathfrak{L} | x\mathfrak{N} \subset \mathfrak{C}^{i-1}_{\mathfrak{L}}(\mathfrak{N})\}.$$

Obviously, $\mathfrak{C}^i(\mathfrak{L}) = \mathfrak{C}^i_{\mathfrak{L}}(\mathfrak{L})$.

3.2.5 Theorem

Let $\mathfrak{N}$ be a subalgebra of $\mathfrak{L}$ such that $\mathrm{ad}_{\mathfrak{L}}\,\mathfrak{N}$ consists of nilpotent transformations. Then $\mathfrak{C}^i_{\mathfrak{L}}(\mathfrak{N}) = \mathfrak{L}$ for some i.

PROOF. We prove this by induction on dim $\mathfrak{N}$. If dim $\mathfrak{N} \leq 1$, it is trivial. Thus, assume that dim $\mathfrak{N} > 1$. Let $\mathfrak{H}$ be a maximal proper subalgebra of

$\mathfrak{N}$. Then, by induction, there exists a positive integer m such that $\mathfrak{C}^m_{\mathfrak{L}}(\mathfrak{H}) = \mathfrak{L}$ and $\mathfrak{C}^m_{\mathfrak{N}}(\mathfrak{H}) = \mathfrak{N}$. Choose j such that $\mathfrak{C}^j_{\mathfrak{N}}(\mathfrak{H}) \subset \mathfrak{H}$ and $\mathfrak{C}^{j+1}_{\mathfrak{N}}(\mathfrak{H}) \not\subset \mathfrak{H}$, and let $x \in \mathfrak{C}^{j+1}_{\mathfrak{N}}(\mathfrak{H}) - \mathfrak{H}$. Then $kx \oplus \mathfrak{H}$ is a subalgebra of $\mathfrak{N}$ and $x\mathfrak{H} \subset \mathfrak{H}$. Thus, $\mathfrak{N} = kx + \mathfrak{H}$, by the maximality of $\mathfrak{H}$, and it follows that $\mathfrak{H}$ is an ideal of $\mathfrak{N}$. Since $x\mathfrak{H} \subset \mathfrak{H}$, the $\mathfrak{C}^i_{\mathfrak{L}}(\mathfrak{H})$ are ad x-stable, as we now show by induction on i. Thus, $\mathfrak{C}^0_{\mathfrak{L}}(\mathfrak{H}) = \{0\}$ is ad x-stable and, if $\mathfrak{C}^{i-1}_{\mathfrak{L}}(\mathfrak{H})$ is ad x-stable and if $y \in \mathfrak{C}^i_{\mathfrak{L}}(\mathfrak{H})$ and $h \in \mathfrak{H}$, then $(yx)h = (yh)x + y(xh) \in \mathfrak{C}^{i-1}_{\mathfrak{L}}(\mathfrak{H})x + \mathfrak{C}^i_{\mathfrak{L}}(\mathfrak{H})(xh) \subset \mathfrak{C}^{i-1}_{\mathfrak{L}}(\mathfrak{H})$, or $(y \operatorname{ad} x)\mathfrak{H} \subset \mathfrak{C}^{i-1}_{\mathfrak{L}}(\mathfrak{H})$ and $y \operatorname{ad} x \in \mathfrak{C}^i_{\mathfrak{L}}(\mathfrak{H})$. Finally, since ad x is nilpotent and stabilizes the spaces in the chain $\{0\} = \mathfrak{C}^0_{\mathfrak{L}}(\mathfrak{H}) \subset \ldots \subset \mathfrak{C}^m_{\mathfrak{L}}(\mathfrak{H}) = \mathfrak{L}$, this chain has a refinement $\{0\} = \mathfrak{V}^0 \subset \ldots \subset \mathfrak{V}^n = \mathfrak{L}$ such that $\mathfrak{V}^i x \subset \mathfrak{V}^{i-1}$. One automatically has $\mathfrak{V}^i\mathfrak{H} \subset \mathfrak{V}^{i-1}$ since $\{\mathfrak{V}^i\}$ refines $(\mathfrak{C}^i_{\mathfrak{L}}(\mathfrak{H}))$ and $\mathfrak{C}^i_{\mathfrak{L}}(\mathfrak{H})\mathfrak{H} \subset \mathfrak{C}^{i-1}_{\mathfrak{L}}(\mathfrak{H})$. It follows that $\mathfrak{V}^i\mathfrak{N} \subset \mathfrak{V}^{i-1}$ for $1 \leq i \leq n$. Thus, $\mathfrak{C}^n_{\mathfrak{L}}(\mathfrak{N}) = \mathfrak{L}$.

3.2.6 Theorem (Engel)

$\mathfrak{L}$ is nilpotent iff ad $\mathfrak{L}$ consists of nilpotent transformations.

PROOF. If $\mathfrak{L}^i = \{0\}$, then $(\operatorname{ad} y)^{i-1} = 0$ for $y \in \mathfrak{L}$. Conversely, suppose that ad $\mathfrak{L}$ consists of nilpotent transformations. Then $\mathfrak{L} = \mathfrak{C}^i_{\mathfrak{L}}(\mathfrak{L}) = \mathfrak{C}^i(\mathfrak{L})$ for some i. Thus $\mathfrak{L}$ is nilpotent by 3.2.3.

3.2.7 Corollary

A maximal proper subalgebra $\mathfrak{H}$ of a nilpotent Lie algebra $\mathfrak{L}$ is an ideal of codimension 1.

PROOF. Take $\mathfrak{N} = \mathfrak{L}$ in the proof of 3.2.5. The proof then shows that $\mathfrak{H}$ is an ideal of $\mathfrak{L}$ of codimension 1.

3.2.8 Definition

Let $\mathfrak{N}$ be a set, $\mathfrak{V}$ an $\mathfrak{N}$-module. Then $\mathfrak{V}^i_0(\mathfrak{N})$ is defined recursively by $\mathfrak{V}^0_0(\mathfrak{N}) = \{0\}$, $\mathfrak{V}^i_0(\mathfrak{N}) = \{v \in \mathfrak{V} \mid v\mathfrak{N} \subset \mathfrak{V}^{i-1}_0(\mathfrak{N})\}$.

If $\mathfrak{N}$ is a subalgebra of $\mathfrak{L}$ and $\mathfrak{V} = \mathfrak{L}$, then $\mathfrak{V}^i_0(\operatorname{ad} \mathfrak{N}) = \mathfrak{C}^i_{\mathfrak{L}}(\mathfrak{N})$.

3.2.9 Theorem (Engel)

Let $\mathfrak{N}$ be a subalgebra of $(\operatorname{Hom}_k V)_{\text{Lie}}$, where V is a finite-dimensional vector space over k, consisting of nilpotent transformations. Then $V = V_0^i(\mathfrak{N})$ for some i. Thus, relative to a suitable basis for V, the subalgebra $\mathfrak{N}$ is represented by upper triangular nilpotent matrices.

PROOF. Since $y(\mathrm{ad}\, x)^n$ is a linear combination of terms $x^m y x^{n-m}$ for x, $y \in \mathrm{Hom}_k V$, $x^{2n} = 0$ implies $(\mathrm{ad}\, x)^n = 0$. Thus, $\mathrm{ad}_{\mathfrak{N}} \mathfrak{N}$ consists of nilpotent transformations. Let us regard V as a Lie algebra with product $VV = \{0\}$, and let $\mathfrak{L}$ be the split extension $\mathfrak{L} = \mathfrak{N} \oplus V$ of $\mathfrak{N}$ by V (see 3.4). Then $\mathfrak{N}$ is a subalgebra of $\mathfrak{L}$ and V an ideal of $\mathfrak{L}$. Since $\mathrm{ad}_{\mathfrak{N}} \mathfrak{N}$ and $\mathfrak{N}$ consist of nilpotent transformations, $\mathrm{ad}_{\mathfrak{L}} \mathfrak{N}$ consists of nilpotent transformations. Thus, $\mathfrak{L} = \mathfrak{C}_{\mathfrak{L}}^i(\mathfrak{N}) = \mathfrak{L}_0^i(\mathfrak{N})$ for some i. Thus, $V = V_0^i(\mathfrak{N})$ for some i since $V \subset \mathfrak{L}$.

3.2.10 Theorem

Let f be a representation of $\mathfrak{L}$ on a finite-dimensional nonzero vector space V over k. Suppose that $f(x)$ is nilpotent for $x \in \mathfrak{L}$. Then there exists a nonzero vector $v \in V$ such that $vf(x) = 0$ for all $x \in \mathfrak{L}$.

PROOF. Take any nonzero vector in $V_0^1(\mathfrak{N})$. This exists since $V_0^i(\mathfrak{N}) = V$ for some i.

3.2.11 Proposition

Let $\mathfrak{V}$ be a finite-dimensional $\mathfrak{L}$-module over k. Suppose that $\mathfrak{L}$ is nilpotent and that each $x \in \mathfrak{L}$ is split over k with respect to $\mathfrak{V}$. Then $\mathfrak{V} = \sum V_a(\mathfrak{L})$ (a ranges over all functions $a : \mathfrak{L} \to k$).

PROOF. The proof is by induction on $\dim \mathfrak{V}$. We may take $\mathfrak{L}$ to be a subalgebra of $(\mathrm{Hom}_k \mathfrak{V})_{\mathrm{Lie}}$. For $x \in \mathfrak{L}$, $\mathrm{ad}_{\mathfrak{L}}\, x$ is nilpotent. Thus, $0 = (\mathrm{ad}_{\mathfrak{L}}\, x)_s = \mathrm{ad}_{\mathfrak{L}}\, x_s$ and $[y, x_s] = 0$ for x, $y \in \mathfrak{L}$. We therefore have that $S = \{x_s \mid x \in \mathfrak{L}\}$ is a family of linear transformations of $\mathfrak{V}$ each of which commutes with all elements of $\mathfrak{L}$. If each $x_s \in S$ acts on $\mathfrak{V}$ through a single scalar $a(x)$, then $\mathfrak{V} = V_a(\mathfrak{L})$ where a is the function $x \mapsto a(x)$. Thus, we may assume that there exist $x_s \in S$ and distinct scalars $\alpha_1, \ldots, \alpha_m$ with $m > 1$ such that $\mathfrak{V} = \sum \oplus V_{\alpha_i}(x_s)$ and $\mathfrak{V}_{\alpha_i}(x_s) \neq \{0\}$ for $1 \leq i \leq m$. By the commutativity, the $\mathfrak{V}_{\alpha_i}(x_s) = \mathfrak{V}_i$ are $\mathfrak{L}$-stable. Applying induction to each $\mathfrak{V}_i$, we have $\mathfrak{V}_i = \sum_a V_{ia}(\mathfrak{L})$. Thus, $\mathfrak{V} = \sum_a V_a(\mathfrak{L})$. Since $\mathfrak{V}_i = V_{\alpha_i}(x_s) = V_{\alpha_i}(x)$ and since the α_i are distinct, we have $\mathfrak{V}_a(\mathfrak{L}) \neq \{0\} \Rightarrow a(x) = \alpha_i$ for some $i \Rightarrow \mathfrak{V}_a(\mathfrak{L}) \subset \mathfrak{V}_i$ for some i. Since the sums $\sum \mathfrak{V}_i$ and $\sum \mathfrak{V}_{ia}(\mathfrak{L})$ are direct, it follows that $\mathfrak{V} = \sum \mathfrak{V}_a(\mathfrak{L})$ is direct.

3.2.12 Definition

If $\mathfrak{V}_a(\mathfrak{N}) \neq \{0\}$, $\mathfrak{V}_a(\mathfrak{N})$ is called the *weight space* for $\mathfrak{N}$ in $\mathfrak{V}$ with respect to a and a is called a *weight* of $\mathfrak{N}$ in $\mathfrak{V}$.

Since $\mathfrak{V}$ is finite dimensional, $\mathfrak{N}$ has only finitely many weights in $\mathfrak{V}$.

3.2.13 Theorem

Let $\mathfrak{N}$ be a nilpotent subalgebra of $\mathfrak{L}$ and let $\mathfrak{V}$ be a finite-dimensional $\mathfrak{L}$-module. Regard $\mathfrak{L}$ as an $\mathfrak{N}$-module via ad. Then $\mathfrak{V}_a(\mathfrak{N})\mathfrak{L}_b(\mathfrak{N}) \subset \mathfrak{V}_{a+b}(\mathfrak{N})$ for all functions $a, b : \mathfrak{N} \to k$.

PROOF. Consider the split extension $\mathfrak{B} = \mathfrak{L} \oplus \mathfrak{V}$ of $\mathfrak{L}$ by $\mathfrak{V}$, $\mathfrak{V}$ having the product $\mathfrak{V}\mathfrak{V} = \{0\}$. Then $\operatorname{ad}_{\mathfrak{B}} \mathfrak{N} \subset \operatorname{Der} \mathfrak{B}$ and $\mathfrak{B}_a(\operatorname{ad} \mathfrak{N}) = \mathfrak{L}_{a'}(\mathfrak{N}) \oplus \mathfrak{V}_{a'}(\mathfrak{N})$ where $a' = a \circ \operatorname{ad}$. Since $\mathfrak{B}_a(\operatorname{ad} \mathfrak{N})\mathfrak{B}_b(\operatorname{ad} \mathfrak{N}) \subset \mathfrak{B}_{a+b}(\operatorname{ad} \mathfrak{N})$, by 2.4.12, and since $\mathfrak{V}$ is an ideal of $\mathfrak{B}$, $\mathfrak{V}_{a'}(\mathfrak{N})\mathfrak{L}_{b'}(\mathfrak{N}) \subset \mathfrak{V}_{a'+b'}(\mathfrak{N})$ for all a', b'.

3.2.14 Corollary

Let $\mathfrak{N}$ be a nilpotent ideal in $\mathfrak{L}$, and let $\mathfrak{V}$ be a finite-dimensional $\mathfrak{L}$-module. Then $\mathfrak{V}_a(\mathfrak{N})$ is an $\mathfrak{L}$-submodule of $\mathfrak{V}$ for every $a : \mathfrak{N} \to k$.

PROOF. $\mathfrak{L} = \mathfrak{L}_0(\operatorname{ad} \mathfrak{N})$ since $\mathfrak{N}$ is a nilpotent ideal. Thus, $\mathfrak{V}_a(\mathfrak{N})\mathfrak{L} \subset \mathfrak{V}_{a+0}(\mathfrak{N}) = \mathfrak{V}_a(\mathfrak{N})$ for all a.

3.3 Cartan Subalgebras

In this section, k is infinite of arbitrary characteristic.

The Cartan subalgebras of a Lie algebra $\mathfrak{L}$ are certain nilpotent subalgebras that are particularly well suited as objects by way of which to study $\mathfrak{L}$. We show here that they exist and that the decomposition of $\mathfrak{L}$ into weight spaces for a given split Cartan subalgebra $\mathfrak{H}$ provides a rough description of a multiplication table for $\mathfrak{L}$.

3.3.1 Definition

A subalgebra $\mathfrak{H}$ of $\mathfrak{L}$ is a *Cartan subalgebra* of $\mathfrak{L}$ if $\mathfrak{H}$ is nilpotent and $\mathfrak{H} = \mathfrak{L}_0(\text{ad } \mathfrak{H})$.

If $\mathfrak{H}$ is a Cartan subalgebra of $\mathfrak{L}$, $\mathfrak{N}_{\mathfrak{L}}(\mathfrak{H}) = \mathfrak{H}$ and $\mathfrak{H} \supset \mathfrak{C}_{\mathfrak{L}}(\mathfrak{H})$. For $(\text{ad } x|_{\mathfrak{H}})^n = 0$ implies for $x \in \mathfrak{H}$ that $(\text{ad } x|_{\mathfrak{N}_{\mathfrak{L}}(\mathfrak{H})})^{n+1} = 0$ and $\mathfrak{H} = \mathfrak{L}_0(\text{ad } \mathfrak{H}) \supset \mathfrak{N}_{\mathfrak{L}}(\mathfrak{H}) \supset \mathfrak{C}_{\mathfrak{L}}(\mathfrak{H})$.

3.3.2 Definition

An element x of $\mathfrak{L}$ is *regular* if $\dim \mathfrak{L}_0(\text{ad } x) \leq \dim \mathfrak{L}_0(\text{ad } y)$ for all $y \in \mathfrak{L}$. The set of regular elements of $\mathfrak{L}$ is denoted $\mathfrak{L}_{\text{reg}}$.

3.3.3 Theorem

Let $x \in \mathfrak{L}_{\text{reg}}$. Then $\mathfrak{H} = \mathfrak{L}_0(\text{ad } x)$ is a Cartan subalgebra of $\mathfrak{L}$.

PROOF. We use the Zariski topology. For the basic material on the Zariski topology needed here, see the Appendix. Consider $\mathfrak{L} = \mathfrak{L}_0(\text{ad } x) \oplus \mathfrak{L}_*(\text{ad } x) = \mathfrak{H} \oplus \mathfrak{L}_*(\text{ad } x)$. Then $\mathfrak{L}_*(\text{ad } x)\text{ad } \mathfrak{H} \subset \mathfrak{L}_*(\text{ad } x)$ by 3.4.12. Let $f(y) = \det (\text{ad } y\,|_{\mathfrak{L}_*(\text{ad } x)})$ for $y \in \mathfrak{H}$ and $\mathfrak{H}_f = \{y \in \mathfrak{H} | f(y) \neq 0\}$. Then $\mathfrak{H}_f$ is a Zariski open subset of $\mathfrak{H}$ containing x. It is therefore dense in $\mathfrak{H}$. Now $\mathfrak{L}_0(\text{ad } y) \subset \mathfrak{L}_0(\text{ad } x)$ for $y \in \mathfrak{H}_f$, since ad y stabilizes $\mathfrak{L}_0(\text{ad } x)$ and $\mathfrak{L}_*(\text{ad } x)$, and $\text{ad } y|_{\mathfrak{L}_*(\text{ad } x)}$ is nonsingular. Thus, $\mathfrak{L}_0(\text{ad } y) = \mathfrak{L}_0(\text{ad } x) = \mathfrak{H}$ for $y \in \mathfrak{H}_f$, by the minimality of $\dim \mathfrak{L}_0(\text{ad } x)$. Taking $n = \dim \mathfrak{H}$, we have $(\text{ad } y|_{\mathfrak{H}})^n = 0$ for $y \in \mathfrak{H}_f$. Since $\mathfrak{H}_f$ is dense in $\mathfrak{H}$ and $y \mapsto \text{ad } y|_{\mathfrak{H}}$ is continuous, it follows that $(\text{ad } y|_{\mathfrak{H}})^n = 0$ for all $y \in \mathfrak{H}$. Thus, $\mathfrak{H} = \mathfrak{H}_0(\text{ad } \mathfrak{H})$ and $\mathfrak{H}$ is nilpotent by Engel's theorem. Since $x \in \mathfrak{H}$, $\mathfrak{H} \subset \mathfrak{L}_0(\text{ad } \mathfrak{H}) \subset \mathfrak{L}_0(\text{ad } x) = \mathfrak{H}$ and $\mathfrak{H} = \mathfrak{L}_0(\text{ad } \mathfrak{H})$. Thus, $\mathfrak{H}$ is a Cartan subalgebra of $\mathfrak{L}$.

We now let $\mathfrak{H}$ be a Cartan subalgebra of $\mathfrak{L}$ and regard $\mathfrak{L}$ as an $\mathfrak{H}$-module via the adjoint representation.

3.3.4 Definition

$\mathfrak{H}$ is *split* over k if $\text{ad}_{\mathfrak{L}} x$ is split over k (see 1.4.3) for $x \in \mathfrak{H}$.

3.3.5 Definition

$\mathfrak{L}$ is *split* over k if $\mathfrak{L}$ has a Cartan subalgebra split over k.

3.3.6 Definition

A *root* of $\mathfrak{H}$ in $\mathfrak{L}$ is a function $a : \mathfrak{H} \to k$ such that $\mathfrak{L}_a(\mathfrak{H}) \neq \{0\}$.

Thus, the roots of $\mathfrak{H}$ in $\mathfrak{L}$ are the weights of $\mathfrak{H}$ in $\mathfrak{L}$ where $\mathfrak{L}$ is regarded as an $\mathfrak{H}$-module via ad.

3.3.7 Theorem

Let $\mathfrak{H}$ be a Cartan subalgebra of $\mathfrak{L}$. Then

1. $\mathfrak{L}_a(\mathfrak{H})\mathfrak{L}_b(\mathfrak{H}) \subset \mathfrak{L}_{a+b}(\mathfrak{H})$ for all $a, b : \mathfrak{H} \to k$;
2. $\mathfrak{H}\mathfrak{L}_a(\mathfrak{H}) \subset \mathfrak{L}_a(\mathfrak{H})$ for all $a : \mathfrak{H} \to k$;
3. $\mathfrak{L}_a(\mathfrak{H}) \perp \mathfrak{L}_b(\mathfrak{H})$ with respect to $\mathrm{K}(\ ,\ )$ where $a, b : \mathfrak{H} \to k$ are such that $a + b \not\equiv 0$;
4. if $\mathfrak{H}$ is split over k, then $\mathfrak{L} = \sum \mathfrak{L}_a(\mathfrak{H})$ where a ranges over the roots of $\mathfrak{H}$ in L;
5. $\mathfrak{H} \perp \mathfrak{L}_*(\mathrm{ad}\ \mathfrak{H})$ with respect to $\mathrm{K}(\ ,\ )$.

PROOF. All is clear from earlier material, except 3 and 5. Let k' be the algebraic closure of k, $\mathfrak{L}' = \mathfrak{L}_{k'}$. Then $\mathfrak{L}' = \sum \oplus \mathfrak{L}'_{a'}(\mathfrak{H})$ where a' ranges over the functions from $\mathfrak{H}$ into k'. Now let $a' + b' \not\equiv 0$, $x \in \mathfrak{L}'_{a'}(\mathfrak{H})$, $y \in \mathfrak{L}'_{b'}(\mathfrak{H})$. For any c', $\mathfrak{L}'_{c'}(\mathfrak{H})\mathrm{ad}\ x\ \mathrm{ad}\ y \subset \mathfrak{L}'_{c'+a'+b'}(\mathfrak{H})$. If we choose a basis compatible with the decomposition $\mathfrak{L}' = \sum \oplus \mathfrak{L}'_{c'}(\mathfrak{H})$, the matrix of $\mathrm{ad}\ x\ \mathrm{ad}\ y$ is then of the form:

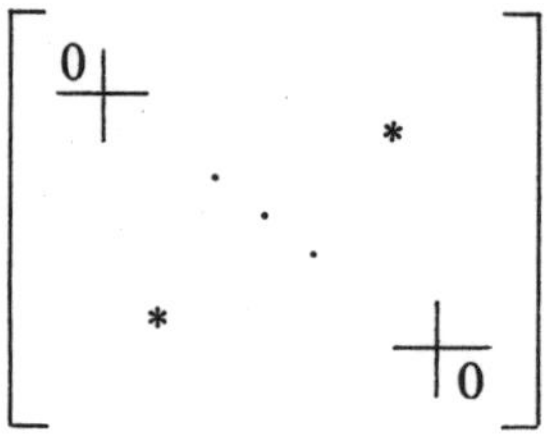

Figure 2.

Thus, $\mathrm{K}(x, y) = 0$ and $\mathfrak{L}'_{a'}(\mathfrak{H}) \perp \mathfrak{L}'_{b'}(\mathfrak{H})$. In particular, $\mathfrak{L}'_0(\mathfrak{H}) \perp \mathfrak{L}'_*(\mathfrak{H})$ since $\mathfrak{L}'_*(\mathfrak{H}) = \sum_{a' \neq 0} L'_{a'}(\mathfrak{H})$. Now $\mathfrak{L}_a(\mathfrak{H})$ is a k-form of $\mathfrak{L}'_a(\mathfrak{H})$ for $a(\mathfrak{H}) \subset k$, $\mathfrak{L}_*(\mathfrak{H})$ is a k-form of $\mathfrak{L}'_*(\mathfrak{H})$, and $\mathfrak{H} = \mathfrak{L}_0(\mathfrak{H})$ is a k-form of $\mathfrak{L}'_0(\mathfrak{H})$. Thus, $\mathfrak{L}_a(\mathfrak{H}) \perp \mathfrak{L}_b(\mathfrak{H})$ for $a + b \neq 0$ and $\mathfrak{H} \perp \mathfrak{L}_*(\mathfrak{H})$.

3.4 Solvable Lie Algebras

Let $\mathfrak{V}$ be a finite-dimensional Lie module for $\mathfrak{L}$ over k. It is necessary to assume that the characteristic of k is 0.

3.4.1 Theorem

Let $h \in \mathfrak{L}$ and suppose that $h \in (\mathfrak{L}_0(\text{ad } h))^{(1)}$. Then $\mathfrak{V} = \mathfrak{V}_0(h)$.

PROOF. The hypothesis is preserved under ascent, the conclusion under descent. Thus, we may assume without loss of generality that the ground field is algebraically closed. Now let $h = \sum x_i y_i$ where x_i, $y_i \in \mathfrak{L}_0(\text{ad } h)$.

Then $\mathfrak{V} = \sum \mathfrak{V}_\alpha(h)$ and each $\mathfrak{V}_\alpha(h)$ is $\mathfrak{L}_0(\text{ad } h)$-stable, by 3.2.13, hence stable under the x_i,y_i. Thus, letting f: $\mathfrak{L} \to (\text{Hom}_k\ \mathfrak{V})_{\text{Lie}}$ be the representation of $\mathfrak{L}$ afforded by $\mathfrak{V}$, $f(h)|_{\mathfrak{V}_\alpha(h)} = \sum f(x_i y_i)|_{\mathfrak{V}_\alpha(h)} = \sum [f(x_i), f(y_i)]|_{\mathfrak{V}_\alpha(h)}$. Consequently, Trace $f(h)|_{\mathfrak{V}_\alpha(h)} = \sum$ Trace $[f(x_i), f(y_i)]|_{\mathfrak{V}_\alpha(h)} = 0$, since the trace of an element of the form $xy - yx$ is 0. But Trace $f(h)|_{\mathfrak{V}_\alpha(h)} = \alpha \dim \mathfrak{V}_\alpha(h)$. Thus, $\alpha = 0$ if $\mathfrak{V}_\alpha(h) \neq 0$ and $\mathfrak{V} = \sum \mathfrak{V}_\alpha(h) = \mathfrak{V}_0(h)$.

3.4.2 Corollary

If ad h is nilpotent and $h \in \mathfrak{L}^{(1)}$, then $\mathfrak{V} = \mathfrak{V}_0(h)$.

3.4.3 Theorem

Let $\mathfrak{L}$ be solvable. Then $\mathfrak{V} = \mathfrak{V}_0(\mathfrak{L}^{(1)})$.

PROOF. Suppose not and take a counterexample with $\dim \mathfrak{L} + \dim \mathfrak{V}$ minimal. Choose n maximal such that $\mathfrak{L}^{(n)} \neq 0$, and let $\mathfrak{A} = \mathfrak{L}^{(n)}$. Then $\mathfrak{A} = \mathfrak{L}^{(n)}$ is Abelian and, since $\mathfrak{V} \neq \mathfrak{V}_0(\mathfrak{L}^{(1)})$, $n \geq 1$. Now $\mathfrak{A} = \mathfrak{L}^{(n)}$ is spanned by products xy with $x, y \in \mathfrak{L}^{(n-1)}$. For such an xy, $\mathfrak{L} = \mathfrak{L}_0(\text{ad } xy)$ since $\mathfrak{A}$ is an Abelian ideal of $\mathfrak{L}$. Thus, $\mathfrak{V} = \mathfrak{V}_0(xy)$ by 3.4.2. Thus, $\mathfrak{V} = \mathfrak{V}_0(\mathfrak{A})$ since $\mathfrak{A}$ is Abelian. Let $\mathfrak{W} = \{v \in \mathfrak{V} \mid v\mathfrak{A} = 0\}$. Since $\mathfrak{V}_0(\mathfrak{A}) = \mathfrak{V} \neq \{0\}$, $\mathfrak{W} \neq \{0\}$. If $\mathfrak{W} = \mathfrak{V}$, then $\mathfrak{V}\mathfrak{A} = 0$ and $\mathfrak{V}$ may be regarded as an $\mathfrak{L}/\mathfrak{A}$-module. But then $\mathfrak{V} = \mathfrak{V}_0((\mathfrak{L}/\mathfrak{A})^{(1)})$ by the minimality of $\dim \mathfrak{V} + \dim \mathfrak{L}$, so $\mathfrak{V} = \mathfrak{V}_0(\mathfrak{L}^{(1)})$, a contradiction. Thus, $\mathfrak{W} \subsetneq \mathfrak{V}$. Then $\mathfrak{W}$ and $\mathfrak{V}/\mathfrak{W}$ may be regarded as Lie modules for $\mathfrak{L}$, since clearly $\mathfrak{W}\mathfrak{L} \subset \mathfrak{W}$. Now $\mathfrak{W} = \mathfrak{W}_0(\mathfrak{L}^{(1)})$ and $\mathfrak{V}/\mathfrak{W} = (\mathfrak{V}/\mathfrak{W})_0(\mathfrak{L}^{(1)})$ by the minimality of $\dim \mathfrak{L} + \dim \mathfrak{V}$. Thus, $V = \mathfrak{V}_0(\mathfrak{L}^{(1)})$, a contradiction.

3.4.4 Corollary

Let $\mathfrak{L}$ be solvable. Then the set $\mathfrak{N} = \{x \in \mathfrak{L} \mid \mathfrak{V} = \mathfrak{V}_0(x)\}$ is an ideal of $\mathfrak{L}$ containing $\mathfrak{L}^{(1)}$.

PROOF. We have seen that $\mathfrak{N} \supset \mathfrak{L}^{(1)}$. Now let $\mathfrak{H}$ be a maximal subalgebra of $\mathfrak{L}$ contained in $\mathfrak{N}$ and containing $\mathfrak{L}^{(1)}$ such that $\mathfrak{V} = \mathfrak{V}_0(\mathfrak{H})$. We claim that $\mathfrak{N} = \mathfrak{H}$, whence $\mathfrak{N}$ is an ideal of $\mathfrak{L}$ since $\mathfrak{N}\mathfrak{L} \subset \mathfrak{L}^{(1)} \subset \mathfrak{N}$. Thus, let $x \in \mathfrak{N}$. Since $\mathfrak{H}\text{ad } x \subset \mathfrak{L}^{(1)} \subset \mathfrak{H}$, the terms of the series $\{0\} = \mathfrak{V}_0^0(\mathfrak{H}) \subset \ldots \subset \mathfrak{V}_0^m(\mathfrak{H}) = \mathfrak{V}$ are x-stable. Since $\mathfrak{V} = \mathfrak{V}_0(x)$, there is a refinement $\{0\} = \mathfrak{V}^0 \subset \ldots \subset \mathfrak{V}^n = \mathfrak{V}$ which is x-stable and $\mathfrak{H}$-stable, with $\mathfrak{V}^i(kx + \mathfrak{H}) \subset \mathfrak{V}^{i-1}$ for all i. Thus, $\mathfrak{V} = \mathfrak{V}_0(kx + \mathfrak{H})$. By the maximality of $\mathfrak{H}$, $kx + \mathfrak{H} = \mathfrak{H}$ and $x \in \mathfrak{H}$. Thus, $\mathfrak{H} = \mathfrak{N}$.

3.4.5 Corollary

Let $\mathfrak{L}$ be a subalgebra of $(\mathrm{Hom}_k\ \mathfrak{V})_{\mathrm{Lie}}$, where $\mathfrak{V}$ is finite dimensional over k. If $\mathfrak{R}$ is a solvable ideal of $\mathfrak{L}$, then the set $\mathfrak{N}$ of nilpotent elements of $\mathfrak{R}$ is an ideal of $\mathfrak{L}$ and $[\mathfrak{L}, \mathfrak{R}] \subset \mathfrak{N}$. If $\mathfrak{R}$ is a nilpotent ideal of $\mathfrak{L}$, then $\mathfrak{L}^{(1)} \cap \mathfrak{R}$ consists of nilpotent elements.

PROOF. Suppose that $\mathfrak{R}$ is a solvable ideal of $\mathfrak{L}$. Let $x \in \mathfrak{L}$ and $\mathfrak{S} = kx + \mathfrak{R}$. Then $\mathfrak{S}$ is solvable and $\mathfrak{S}^{(1)}$, hence $[x, \mathfrak{R}]$, is contained in the set $\mathfrak{N}$ of nilpotent elements of $\mathfrak{R}$ by 3.4.4. Since this is true for any $x \in \mathfrak{L}$, $[\mathfrak{L}, \mathfrak{R}] \subset \mathfrak{N}$. By 3.4.4, it follows that $\mathfrak{N}$ is an ideal of $\mathfrak{L}$.

Now let $\mathfrak{R}$ be a nilpotent ideal of $\mathfrak{L}$ and let $h \in \mathfrak{L}^{(1)} \cap \mathfrak{R}$. Then $\mathfrak{L} = \mathfrak{L}_0(\mathrm{ad}\ h)$ since $\mathfrak{R}$ is a nilpotent ideal of $\mathfrak{L}$. Thus, h is nilpotent by 3.4.2.

3.4.6 Corollary

Let $\mathfrak{L}$ be a subalgebra of $(\mathrm{Hom}_k\ \mathfrak{V})_{\mathrm{Lie}}$, $\mathfrak{V}$ finite dimensional over k. Suppose that $\mathfrak{V}$ is $\mathfrak{L}$-completely reducible. Then $\mathfrak{R} = \mathrm{Rad}\ \mathfrak{L}$ is central in $\mathfrak{L}$ and $\mathfrak{L}^{(1)} \cap \mathrm{Rad}\ \mathfrak{L} = \{0\}$. (Thus, by Levi's theorem of 3.6, $\mathfrak{L} = \mathfrak{L}^{(1)} \oplus \mathrm{Rad}\ \mathfrak{L}$ and $\mathfrak{L}^{(1)}$ is semisimple.)

PROOF. Let $\mathfrak{N}$ be the ideal of nilpotent elements of $\mathfrak{R}$. Let $\mathfrak{V} = \sum \oplus \mathfrak{V}_i$ where the $\mathfrak{V}_i$ are $\mathfrak{L}$-irreducible. Since $\mathfrak{N}$ is a nilpotent ideal of $\mathfrak{L}$, $(\mathfrak{V}_i)_0^1$ is a nonzero $\mathfrak{L}$-submodule of $\mathfrak{V}_i$ and $(\mathfrak{V}_i)_0^1 = \mathfrak{V}_i$ by the $\mathfrak{L}$-irreducibility of

$\mathfrak{V}_i$. (For this, see 3.2.9 and 3.2.14.) Thus, $\mathfrak{N} = \{0\}$. Since $[\mathfrak{L}, \mathfrak{N}] \subset \mathfrak{N} = \{0\}$, $\mathfrak{N}$ is central. Now $\mathfrak{N}$ is Abelian, so $\mathfrak{L}^{(1)} \cap \mathfrak{N} \subset \mathfrak{N} = \{0\}$, by 3.4.5. Thus, $\mathfrak{L}^{(1)} \cap \mathfrak{N} = \{0\}$.

3.4.7 Theorem (Lie)
Let $\mathfrak{L}$ be a solvable Lie algebra over k, and let $\mathfrak{V}$ be a finite-dimensional irreducible Lie module for $\mathfrak{L}$ over k. Suppose that $\mathfrak{L}$ has a Cartan subalgebra $\mathfrak{H}$ such that every $x \in \mathfrak{H}$ is split on $\mathfrak{V}$ over k. Then dim $\mathfrak{V} \leq 1$.

PROOF. Assume that $\mathfrak{V} \neq \{0\}$. Then $\mathfrak{W} = \mathfrak{V}_0^1(\mathfrak{L}^{(1)})$ is nonzero by 3.4.3. Now $\mathfrak{W}$ is an $\mathfrak{L}$-submodule, so $\mathfrak{W} = \mathfrak{V}$, by the $\mathfrak{L}$-irreducibility of $\mathfrak{V}$. Thus, $\mathfrak{V}\mathfrak{L}^{(1)} = \{0\}$ and we may regard $\mathfrak{V}$ as an $\mathfrak{A}$-module, $\mathfrak{A} = \mathfrak{L}/\mathfrak{L}^{(1)}$. Now $\mathfrak{A}$ is Abelian and $\mathfrak{L} = \mathfrak{H} \oplus \mathfrak{L}_*(\text{ad } \mathfrak{H})$, so $\mathfrak{L}^{(1)} \supset \mathfrak{L}_*(\text{ad } \mathfrak{H})$ and $\mathfrak{L} = \mathfrak{H} + \mathfrak{L}^{(1)}$. Thus, $\mathfrak{A} = \mathfrak{H} + \mathfrak{L}^{(1)}/\mathfrak{L}^{(1)}$. Since every element of $\mathfrak{H}$ is split over k, every element of $\mathfrak{A}$ is split over k. Thus, $\mathfrak{V} = \sum \oplus \mathfrak{V}_a(\mathfrak{A})$. But $\mathfrak{V}$ is $\mathfrak{A}$-irreducible. This can happen only if dim $\mathfrak{V} = 1$.

The foregoing theorem shows that if $\mathfrak{L}$ is a "split solvable Lie subalgebra" of $(\text{Hom}_k\, \mathfrak{V})_{\text{Lie}}$ and $\mathfrak{V}$ a finite-dimensional vector space over k, then the composition factors of a composition series $\mathfrak{V}_0 = \{0\} \subset \mathfrak{V}_1 \subset \ldots \subset \mathfrak{V}_m = \mathfrak{V}$ of $\mathfrak{V}$ as an $\mathfrak{L}$-module are of dimension 1. Thus, for such $\mathfrak{L}$ and $\mathfrak{V}$, a basis for $\mathfrak{V}$ can be chosen such that the matrix of each element of $\mathfrak{L}$ relative to the chosen basis is in upper triangular form:

$$\begin{bmatrix} \alpha_1 & & * \\ & \ddots & \\ 0 & & \alpha_m \end{bmatrix}$$

Figure 3.

In particular, every element x of such an L is split over k. We refer to this as the *simultaneous triangulability* of a "split solvable subalgebra" of $(\text{Hom}_k\, \mathfrak{V})_{\text{Lie}}$.

3.5 Cartan's Criteria

We now give important criteria for the solvability and the semisimplicity of $\mathfrak{L}$ in terms of the Killing form K (,) of $\mathfrak{L}$ introduced in 2.8. These criteria form the basis for the classification of semisimple Lie algebras which we give in 3.7.

We let $\mathfrak{B}$ be a finite-dimensional Lie module for $\mathfrak{L}$ over k, and f the representation of $\mathfrak{L}$ afforded by $\mathfrak{B}$. The characteristic is 0.

3.5.1 Theorem

Let $xy = h$ with $x \in \mathfrak{L}_{-\alpha}(\operatorname{ad} h)$, $y \in \mathfrak{L}_{\alpha}(\operatorname{ad} h)$ where $\alpha \in k$. Suppose that $\mathfrak{B} \neq \mathfrak{B}_0(h)$. Then Trace $(f(h))^2 \neq 0$ and Trace $f(h)^2$ is a positive rational multiple of α^2. In particular, $\alpha \neq 0$.

PROOF. The hypothesis is preserved under ascent and the conclusion under descent, so we may assume without loss of generality that k is algebraically closed. Now $\mathfrak{B} = \sum \oplus \mathfrak{B}_\beta(h)$. Let β be an eigenvalue of $f(h)$ and $\mathfrak{W}_\beta = \sum_{i \in \mathbb{Z}} V_{\beta + i\alpha}(h)$. This is a subspace stable under x and y, by 3.2.13. Thus, $0 = \text{Trace}\,[f(x), f(y)]|_{\mathfrak{W}_\beta} = \text{Trace}\, f(h)|_{\mathfrak{W}_\beta} = \sum d_i(\beta + i\alpha)$ where $d_i = \dim \mathfrak{B}_{\beta + i\alpha}(h)$. Thus, $\beta = -\dfrac{\sum id_i}{\sum d_i}\alpha$. Thus, each such β has the form $\beta = r_\beta \alpha$ for some rational number r_β. Thus, Trace $f(h)^2 = \sum d_\beta (r_\beta \alpha)^2 = (\sum d_\beta r_\beta^2)\alpha^2$ where $d_\beta = \dim \mathfrak{B}_\beta(h)$. Since $\mathfrak{B} \neq \mathfrak{B}_0(h)$, $\dim \mathfrak{B}_\beta(h) \neq 0$ for some $\beta \neq 0$. Choose such a β. Then $d_\beta \neq 0$, and $r_\beta \neq 0$ and $\alpha \neq 0$ since $\beta = r_\beta \alpha$. Thus, $\sum d_\beta r_\beta^2 > 0$ and Trace $(f(h))^2 = (\sum d_\beta r_\beta^2)\alpha^2 \neq 0$.

3.5.2 Theorem

Let $(x, y) = \text{Trace}\, f(x)f(y)$. Then $\mathfrak{L}$ is solvable iff Kernel f is solvable and $(x, y) = 0$ for $x, y \in \mathfrak{L}^{(1)}$.

PROOF. We may assume without loss of generality that k is algebraically closed. If $\mathfrak{L}$ is solvable, then Kernel $\mathfrak{L}$ is solvable and $f(\mathfrak{L})$ is a solvable subalgebra of $(\operatorname{Hom}_k \mathfrak{B})_{\text{Lie}}$. By Lie's theorem, $f(\mathfrak{L})$ is simultaneously triangulable. Thus, relative to a suitable basis, the elements of $f(\mathfrak{L})^{(1)} = f(\mathfrak{L}^{(1)})$ are nilpotent upper triangular matrices. Thus, $(x, y) = \text{Trace}\, f(x)f(y) = 0$ for $x, y \in \mathfrak{L}^{(1)}$.

We prove the converse by induction on dim $\mathfrak{L}$. If dim $\mathfrak{L} = 1$, the assertion is trivial. It suffices to show that $f(\mathfrak{L})$ is solvable, for then $\mathfrak{L}$/Kernel $\mathfrak{L}$ and Kernel $\mathfrak{L}$ are solvable, so $\mathfrak{L}$ is solvable. Thus, we may assume that $\mathfrak{L}$ is a subalgebra of $(\text{Hom}_k \mathfrak{V})_{\text{Lie}}$ and $f(x) = x$ for $x \in \mathfrak{L}$. Let $\mathfrak{H}$ be a Cartan subalgebra of $\mathfrak{L}$, $\mathfrak{L} = \mathfrak{H} \oplus \sum \mathfrak{L}_a$ where the a's are roots of $\mathfrak{H}$ and $\mathfrak{L}_a = \mathfrak{L}_a(\mathfrak{H})$. Let $x \in \mathfrak{L}_{-a}$, $y \in \mathfrak{L}_a$, $h = [x, y]$. Then $0 = (h, h) = \text{Trace } h^2$ since $h \in \mathfrak{L}^{(1)}$. Thus, $\mathfrak{V} = \mathfrak{V}_0(h)$, by 3.5.1. Thus, $[\mathfrak{L}_{-a}, \mathfrak{L}_a] \subset \mathfrak{N} = \{x \in \mathfrak{H} \mid x \text{ is nilpotent}\}$ for all a. But $\mathfrak{N}$ is an ideal of $\mathfrak{H}$, by 3.4.4. Thus, $\mathfrak{I} = \mathfrak{N} + \sum_{a \neq 0} \mathfrak{L}_a$ is an ideal of $\mathfrak{L}$. If $\mathfrak{I} \subsetneq \mathfrak{L}$, then $\mathfrak{I}$ is solvable, by induction, and $\mathfrak{L}/\mathfrak{I} = \mathfrak{H}/\mathfrak{N}$ is solvable, so that $\mathfrak{L}$ is solvable. Otherwise $\mathfrak{I} = \mathfrak{L}$. Then $\mathfrak{H} = \mathfrak{N}$. But then x is nilpotent for $x \in \mathfrak{H} = \mathfrak{N}$, so that ad x is nilpotent for $x \in \mathfrak{H}$, by 1.4.8. Then $\mathfrak{L} = \mathfrak{L}_0(\text{ad } \mathfrak{H}) = \mathfrak{H}$ and $\mathfrak{L}$ is again solvable.

The above proof would be more natural if we knew that the ideal $\mathfrak{N} + \sum \mathfrak{L}_a$ is nilpotent, for we would not need an auxiliary argument for the case $\mathfrak{I} = \mathfrak{L}$. The nilpotency of $\mathfrak{N} + \sum \mathfrak{L}_a$ follows from the strong version of Engel's theorem given in 4.3.7.

3.5.3 Theorem

If $\mathfrak{L}$ is semisimple, then the trace form $(x, y) = \text{Trace } f(x)f(y)$ is nondegenerate.

PROOF. The ideal $\mathfrak{L}^{\perp}$ is solvable, by 3.5.2. Thus, $\mathfrak{L}^{\perp} = \{0\}$ and $(\ ,\ )$ is nondegenerate.

3.5.4 Corollary

$\mathfrak{L}$ is semisimple iff $K(\ ,\ )$ is nondegenerate.

PROOF. By 2.8.4 and 3.5.3.

3.5.5 Corollary

For $\mathfrak{I}$ an ideal of $\mathfrak{L}$, $\mathfrak{L}/\mathfrak{I}$ is semisimple iff $\mathfrak{I} \supset \text{Rad } \mathfrak{L}$.

PROOF. If $\mathfrak{L}/\mathfrak{I}$ is semisimple, then the solvable ideal Rad $((\mathfrak{L} + \mathfrak{I})/\mathfrak{I})$ is zero and Rad $\mathfrak{L} \subset \mathfrak{J}$. Conversely let $\mathfrak{J} \supset$ Rad $\mathfrak{L}$ and define $\overline{\mathfrak{L}} = \mathfrak{L}/\text{Rad}\,\mathfrak{L}$, $\overline{\mathfrak{J}} = \mathfrak{J}/\text{Rad}\,\mathfrak{L}$. Then $\mathfrak{L}/\mathfrak{J} \simeq \overline{\mathfrak{L}}/\overline{\mathfrak{J}}$. But $\overline{\mathfrak{L}}$ is semisimple, so $\overline{\mathfrak{L}}$ has nondegenerate Killing form. It therefore follows from 2.7.4 or 2.8.5 that $\overline{\mathfrak{L}}/\overline{\mathfrak{J}}$ is semisimple.

3.6 Existence of Complements

We prove here that Rad $\mathfrak{L}$ has a subalgebra complement in $\mathfrak{L}$. This is Levi's theorem. We also prove that finite-dimensional representations of semisimple Lie algebras (of characteristic 0) are completely reducible. The proofs are based on the Casimir operator, which is defined for any Lie algebra having a nondegenerate invariant form.

Throughout the section, $\mathfrak{V}$ is a finite-dimensional Lie module for $\mathfrak{L}$, and f is the representation of $\mathfrak{L}$ afforded by $\mathfrak{V}$. The characteristic is 0.

We begin by considering the situation when $\mathfrak{L}$ has a nondegenerate invariant form (,). Thus, let $e_1, \ldots, e_n$ and $f_1, \ldots, f_n$ be dual bases for $\mathfrak{L}$ with respect to such a form (,).

3.6.1 Definition

The *Casimir operator* for $\mathfrak{L}$ with respect to (,), the e_i's, and the f_i's is the element $T = \sum f(e_i)f(f_i)$ of $\text{Hom}_k\,\mathfrak{V}$.

3.6.2 Lemma

If $u_i, v_i \in \mathfrak{L}$ for $1 \leq i \leq m$ and $\sum(u_i, w)v_i = 0$ for all $w \in \mathfrak{L}$, then $\sum f(u_i)f(v_i) = 0$.

PROOF. Let $u_i = \sum a_{ij}e_j$ with $a_{ij} \in k$. Letting $x_j = \sum_i a_{ij}v_i$, we have $0 = \sum(u_i, w)v_i = \sum_{ij}(a_{ij}e_j, w)v_i = \sum_j(e_j, w)x_j$ for all $w \in \mathfrak{L}$. Taking $w = f_i$, we have $0 = x_i$ from the equation $0 = \sum_j(e_j, f_i)x_j$. Since the x_j are all 0, we have $\sum f(u_i)f(v_i) = \sum f(a_{ij}e_j)f(v_i) = \sum f(e_j)f(x_j) = 0$.

3.6.3 Theorem

Let T be the Casimir operator $\sum f(e_i)f(f_i)$. Then $[T, f(\mathfrak{L})] = \{0\}$.

PROOF. $[T, f(x)] = \sum f(e_i)[f(f_i), f(x)] + \sum[f(e_i), f(x)]f(f_i) = \sum f(e_i)f(f_i x) + \sum f(e_i x)f(f_i)$. To show that this is 0, it suffices by the lemma to show that $0 = \sum(e_i, w)(f_i x) + \sum(e_i x, w)f_i$ for $w \in \mathfrak{L}$. The first term is $(\sum(e_i, w)f_i)x = wx$ and the second, by the associativity of the form, is $\sum(e_i, xw)f_i = xw$. Thus, their sum is $wx + xw = 0$.

We now assume that $\mathfrak{L}$ is a nonzero semisimple Lie algebra and the representation f is faithful. Then the trace form $(x, y) = \operatorname{Tr} f(x)f(y)$ is nondegenerate, by 3.5.3. Let $e_1, \ldots, e_n$ and $f_1, \ldots, f_n$, be dual bases for $\mathfrak{L}$ with respect to $(\,,\,)$ and let T be the corresponding Casimir operator $T = \sum f(e_i)f(f_i)$. Then Trace $T = \dim \mathfrak{L} \neq 0$, since Trace $T = \sum$Trace $f(e_i)f(f_i) = \sum(e_i, f_i) = \sum 1 = n$. This is crucial in the proof of the following "Fitting's Lemma."

3.6.4 Theorem

Let $\mathfrak{L}$ be semisimple, and let $\mathfrak{V}$ be a finite-dimensional Lie module for $\mathfrak{L}$. Then $\mathfrak{V} = \mathfrak{V}_0 \oplus \mathfrak{V}_*$ uniquely, where $\mathfrak{V}_0$ and $\mathfrak{V}_*$ are $\mathfrak{L}$-submodules such that $\mathfrak{V}_0\mathfrak{L} = \{0\}$ and $\mathfrak{V}_*\mathfrak{L} = \mathfrak{V}_*$.

PROOF. The proof is by induction on dim V and is trivial if $\mathfrak{V} = \{0\}$ or $\mathfrak{L} = \{0\}$. We may assume that the affording representation f is faithful and that $\mathfrak{L} \neq \{0\}$. Take $(\,,\,)$, $e_1, \ldots, e_n$, $f_1, \ldots, f_n$, and T as above. Let $\mathfrak{V} = \mathfrak{V}^0 \oplus \mathfrak{V}^*$ be the Fitting decomposition of $\mathfrak{V}$ relative to T, that is, let $\mathfrak{V}^0 = \mathfrak{V}_0(T)$, $\mathfrak{V}^* = \mathfrak{V}_*(T)$. Since Trace $T = \dim \mathfrak{L} \neq 0$, $\mathfrak{V}^0$ is of dimension less than dim $\mathfrak{V}$. Since $[T, f(\mathfrak{L})] = \{0\}$, $\mathfrak{V}^0$ and $\mathfrak{V}^*$ are $\mathfrak{L}$-submodules. By induction, $\mathfrak{V}^0 = \mathfrak{V}_0 \oplus \mathfrak{W}$ where $\mathfrak{V}_0\mathfrak{L} = \{0\}$ and $\mathfrak{W}\mathfrak{L} = \mathfrak{W}$. Since $\mathfrak{V}^*T = \mathfrak{V}^*$, we have $\mathfrak{V}_*\mathfrak{L} = \mathfrak{V}_*$ where $\mathfrak{V}_* = \mathfrak{W} + \mathfrak{V}^*$. Now $\mathfrak{V} = \mathfrak{V}_0 \oplus \mathfrak{V}_*$ is the desired decomposition. The unicity of $\mathfrak{V}_*$ is trivial since $\mathfrak{V}\mathfrak{L} = \mathfrak{V}_0\mathfrak{L} + \mathfrak{V}_*\mathfrak{L} = \mathfrak{V}_*\mathfrak{L} = \mathfrak{V}_*$. The unicity of $\mathfrak{V}_0$ amounts to $\{u \in \mathfrak{V}_* \mid u\mathfrak{L} = 0\}$ being 0. But this is true since $\mathfrak{V}_* = \mathfrak{W} + \mathfrak{V}^*$ and $\{u \in \mathfrak{W} \mid u\mathfrak{L} = 0\}$ is 0 by induction.

3.6.5 Theorem

Let Rad $\mathfrak{L}$ be Abelian and let $\mathfrak{U}$ be an $\mathfrak{L}$-submodule of $\mathfrak{V}$ such that $\mathfrak{U}$ Rad $\mathfrak{L} = \{0\}$ and $\mathfrak{V}$ Rad $\mathfrak{L} \subset \mathfrak{U}$. Then there exists a projection $\pi : \mathfrak{V} \to \mathfrak{U}$ from $\mathfrak{V}$ onto $\mathfrak{U}$ such that $\mathfrak{L} = \mathfrak{C}(\pi) +$ Rad $\mathfrak{L}$ and $\mathfrak{C}(\pi) \cap$ Rad $\mathfrak{L} =$ Kernel $f \cap$ Rad $\mathfrak{L}$ where $\mathfrak{C}(\pi) = \{x \in \mathfrak{L} \mid \pi f(x) = f(x)\pi\}$, f being the representation afforded by $\mathfrak{V}$.

PROOF. Let $\mathfrak{A} = \text{Rad}\,\mathfrak{L}$. Regard $\mathfrak{N} = \{g \in \text{Hom}_k(\mathfrak{V}, \mathfrak{U}) \mid g|_{\mathfrak{U}} \in k\,\text{id}_{\mathfrak{U}}\}$ as a Lie module for $\mathfrak{L}$, where the module operation is $gx = gf(x) - f(x)g$ for $g \in \mathfrak{N}$, $x \in \mathfrak{L}$. Then it suffices to find $\pi \in \mathfrak{N}$ such that $\pi|_{\mathfrak{U}} = \text{id}_{\mathfrak{U}}$ and $\pi\mathfrak{L} = \pi\mathfrak{A}$. For then $\mathfrak{C}(\pi) = \{x \in \mathfrak{L} \mid \pi x = 0\}$ and $\mathfrak{L} = \mathfrak{C}(\pi) + \mathfrak{A}$; and $\mathfrak{C}(\pi) \cap \mathfrak{A} = \text{Kernel}\, f \cap \mathfrak{A}$ since $x \in \mathfrak{C}(\pi) \cap \mathfrak{A}$, $v \in \mathfrak{V}$ implies $vx = (v\pi)x + v(1 - \pi)x = (v\pi)x + (vx)(1 - \pi) \in \mathfrak{U}x + \mathfrak{U}(1 - \pi) = \{0\}$. To find such a π, we let $\mathfrak{M}$ be the $\mathfrak{L}$-submodule $\{f(x) \mid x \in \mathfrak{A}\}$ of $\mathfrak{N}$, noting that $f(x)$ with $x \in \mathfrak{A}$ maps $\mathfrak{V}$ into $\mathfrak{U}$ by the hypothesis on $\mathfrak{A}$. Since $\mathfrak{A}$ is Abelian, $\mathfrak{M}\mathfrak{A} = \{0\}$. Thus, we may regard $\overline{\mathfrak{N}} = \mathfrak{N}/\mathfrak{M}$ as an $\overline{\mathfrak{L}}$-module where $\overline{\mathfrak{L}} = \mathfrak{L}/\mathfrak{A}$. By 3.6.4, $\overline{\mathfrak{N}} = \overline{\mathfrak{N}}_0 \oplus \overline{\mathfrak{N}}_*$ where $\overline{\mathfrak{N}}_0\overline{\mathfrak{L}} = \{0\}$ and $\overline{\mathfrak{N}}_*\overline{\mathfrak{L}} = \overline{\mathfrak{N}}_*$. Thus, $\mathfrak{N} = \mathfrak{N}_0 + \mathfrak{N}\mathfrak{L} + \mathfrak{M}$ where $\mathfrak{N}_0 = \{g \in \mathfrak{N} \mid g\mathfrak{L} \subset \mathfrak{M}\}$. But $\mathfrak{N}\mathfrak{L}$ and $\mathfrak{M}$ are contained in $\mathfrak{M}' = \{g \in \mathfrak{N} \mid g|_{\mathfrak{U}} = 0\}$. Thus, $\mathfrak{N} = \mathfrak{N}_0 + \mathfrak{M}'$. Since $\mathfrak{N} \neq \mathfrak{M}'$, there exists $\pi \in \mathfrak{N}_0$ such that $\pi|_{\mathfrak{U}} = c \cdot \text{id}_{\mathfrak{U}}$ with $c \neq 0$. We may take $c = 1$, so that π is a projection from $\mathfrak{V}$ onto $\mathfrak{U}$. By choice, $\pi\mathfrak{L} \subset \mathfrak{M}$. But for $x \in \mathfrak{A}$, $\pi x = \pi f(x) - f(x)\pi = -f(x)$ since $\mathfrak{V}f(x) \subset \mathfrak{U}$ and $\pi|_{\mathfrak{U}} = \text{id}_{\mathfrak{U}}$. Thus, $\pi\mathfrak{A} = \mathfrak{M}$. Thus, $\mathfrak{M} = \pi\mathfrak{A} \subset \pi\mathfrak{L} \subset \mathfrak{M}$ and $\pi\mathfrak{A} = \pi\mathfrak{L}$ as desired.

3.6.6 Theorem

Let $\mathfrak{L}$ be semisimple. Then $\mathfrak{V}$ is $\mathfrak{L}$-completely reducible.

PROOF. Let $\mathfrak{U}$ be an $\mathfrak{L}$-submodule of $\mathfrak{V}$. Since $\text{Rad}\,\mathfrak{L} = \{0\}$, the above theorem applies and there exists a projection $\pi: \mathfrak{V} \to \mathfrak{U}$ from $\mathfrak{V}$ onto $\mathfrak{U}$ such that $\mathfrak{L} = \mathfrak{C}(\pi)$. Now $\mathfrak{V} = \mathfrak{U} \oplus \mathfrak{U}'$ where $\mathfrak{U}' = \mathfrak{V}(1 - \pi)$, and $\mathfrak{U}'$ is an $\mathfrak{L}$-submodule because $\mathfrak{L} = \mathfrak{C}(\pi)$.

3.6.7 Theorem (Levi)

Let $\mathfrak{L}$ be a Lie algebra of characteristic 0. Then there exists a semisimple subalgebra $\mathfrak{S}$ of $\mathfrak{L}$ such that $\mathfrak{L} = \mathfrak{S} \oplus \text{Rad}\,\mathfrak{L}$.

PROOF. Suppose first that $\mathfrak{A} = \text{Rad}\,\mathfrak{L}$ is Abelian. Let $\mathfrak{C}$ be the center of $\mathfrak{L}$ and regard $\mathfrak{L}$ as a Lie module for $\mathfrak{L}$ via ad. Applying 3.6.5 to $\mathfrak{U} = \mathfrak{A}$ and $\mathfrak{V} = \mathfrak{L}$, there is a projection $\pi: \mathfrak{L} \to \mathfrak{A}$ from $\mathfrak{L}$ onto $\mathfrak{A}$ such that $\mathfrak{L} = \mathfrak{C}(\pi) + \mathfrak{A}$ and $\mathfrak{C}(\pi) \cap \mathfrak{A} = \text{Kernel ad} \cap \mathfrak{A} = \mathfrak{C}$. Letting $\mathfrak{B} = \mathfrak{C}(\pi)$, $\mathfrak{B}/\mathfrak{C} \simeq \mathfrak{L}/\mathfrak{A}$ is semisimple. Regard $\mathfrak{B}$ as a $\mathfrak{B}/\mathfrak{C}$-module with action

$x(y + \mathfrak{C}) = xy$ for $x \in \mathfrak{B}$, $y + \mathfrak{C} \in \mathfrak{B}/\mathfrak{C}$. Then $\mathfrak{B}$ is $\mathfrak{B}/\mathfrak{C}$–completely reducible, by 3.6.6. Let $\mathfrak{S}$ be a $\mathfrak{B}/\mathfrak{C}$-complement of $\mathfrak{C}$ in $\mathfrak{B}$. Then $\mathfrak{B} = \mathfrak{S} \oplus \mathfrak{C}$ and $\mathfrak{L} = \mathfrak{S} \oplus \mathfrak{A}$.

We now drop the assumption that $\mathfrak{A} = \text{Rad}\,\mathfrak{L}$ is Abelian, and prove by induction on $\dim \mathfrak{L}$ that there is a subalgebra $\mathfrak{S}$ such that $\mathfrak{L} = \mathfrak{S} \oplus \mathfrak{A}$. Then $\mathfrak{S}$ is semisimple since $\mathfrak{S} \simeq \mathfrak{L}/\mathfrak{A}$. If $\mathfrak{L} = \{0\}$ or $\mathfrak{A} = \{0\}$, there is nothing to prove. Thus, take $\mathfrak{A} \neq \{0\}$. Since $\text{Rad}\,(\mathfrak{L}/\mathfrak{A}^{(1)})$ is the Abelian ideal $\mathfrak{A}/\mathfrak{A}^{(1)}$, the preceding paragraph applies. Thus, $\mathfrak{L}/\mathfrak{A}^{(1)} = \mathfrak{B}/\mathfrak{A}^{(1)} \oplus \mathfrak{A}/\mathfrak{A}^{(1)}$ where $\mathfrak{B}$ is a subalgebra of $\mathfrak{L}$ containing $\mathfrak{A}^{(1)}$ and $\mathfrak{B}/\mathfrak{A}^{(1)}$ is semisimple. Since $\mathfrak{A} \neq \{0\}$, $\dim \mathfrak{B} < \dim \mathfrak{L}$. Thus, $\mathfrak{B} = \mathfrak{S} + \mathfrak{A}^{(1)}$ for some subalgebra $\mathfrak{S}$, by induction. But then $\mathfrak{L} = \mathfrak{S} \oplus \mathfrak{A}$.

3.7 Classification of Split Semisimple Lie Algebras

3.7.1 Introduction

We now turn to the classification of split semisimple Lie algebras of characteristic 0. This includes the classification of all semisimple Lie algebras over an algebraically closed field of characteristic 0, since they are split automatically.

The classification is carried out in three steps. In 3.7.2, we attach to each split semisimple Lie algebra $\mathfrak{L}$ of characteristic 0 an abstract root system. In 3.7.3, we classify the abstract root systems. In 3.7.4, we show that two split semisimple Lie algebras of characteristic 0 are isomorphic if and only if their root systems are isomorphic, and describe for almost every isomorphism class of root systems the corresponding Lie algebra. We do not describe the so-called exceptional simple split Lie algebras over k. There are only finitely many isomorphism classes of these exceptional Lie algebras, and these are studied extensively in the literature.

3.7.2 The Root System of $\mathfrak{L}$

We now assume that $\mathfrak{L}$ is a split semisimple Lie algebra over k, k being a field of characteristic 0. We recall that the Killing form $\text{K}(\ ,\ )$ of $\mathfrak{L}$ is nondegenerate and that $\text{Der}\,\mathfrak{L} = \text{ad}\,\mathfrak{L}$ (see 2.8.6). It follows that $\text{ad}\,\mathfrak{L}$ contains the Jordan components $(\text{ad}\,x)_s$, $(\text{ad}\,x)_n$ of $\text{ad}\,x$ for $x \in \mathfrak{L}$ (see 2.4.14). Since the center of $\mathfrak{L}$ is trivial, each $x \in \mathfrak{L}$ has a unique decomposition

$x = x_s + x_n$ such that $(\text{ad } x)_s = \text{ad } x_s$, $(\text{ad } x)_n = \text{ad } x_n$. Clearly $x_s x_n = 0$ since $[(\text{ad } x)_s, (\text{ad } x)_n] = 0$. If $x = x_s$, we say that x is *semisimple.* If $x = x_n$, we say that x is *nilpotent.* A *torus* of $\mathfrak{L}$ is an Abelian subalgebra $\mathfrak{T}$ of $\mathfrak{L}$ consisting of semisimple elements.

3.7.2.1 Theorem

The Cartan subalgebras of $\mathfrak{L}$ are the maximal tori of $\mathfrak{L}$. If $\mathfrak{T}$ is a split Cartan subalgebra of $\mathfrak{L}$, then

1. $\mathfrak{L} = \mathfrak{T} + \sum_{\substack{a \in T^* \\ a \neq 0}} \mathfrak{L}_a$ and $\mathfrak{T} = \mathfrak{L}_0$ where $\mathfrak{T}^*$ is the dual space of $\mathfrak{T}$, and

   $\mathfrak{L}_a = \{x \in \mathfrak{L} \mid xt = a(t)x\}$ for $a \in \mathfrak{T}^*$;
2. $\mathfrak{L}_a \mathfrak{L}_b \subset \mathfrak{L}_{a+b}$ for $a, b \in T^*$;
3. $\mathfrak{L}_a \perp \mathfrak{L}_b$ for $a + b \neq 0$;
4. $\mathfrak{L}_{-a}$ and $\mathfrak{L}_a$ are dual relative to $\text{K}(\ ,\ )$ for $a \in \mathfrak{T}^*$;
5. $\text{K}(\ ,\ )$ is nondegenerate on $\mathfrak{T}$.

PROOF. We show first that any maximal torus $\mathfrak{T}$ of $\mathfrak{L}$ is contained in a Cartan subalgebra $\mathfrak{H}$ of $\mathfrak{L}$. We then show that any Cartan subalgebra of $\mathfrak{L}$ is a torus. This establishes that the Cartan subalgebras of $\mathfrak{L}$ are the maximal tori of $\mathfrak{L}$.

Suppose that $\mathfrak{T}$ is a maximal torus of $\mathfrak{L}$. Let $\mathfrak{H}$ be a Cartan subalgebra of $\mathfrak{C}_{\mathfrak{L}}(\mathfrak{T})$. Then $\mathfrak{T} \subset \mathfrak{N}_{\mathfrak{C}}(\mathfrak{H}) = \mathfrak{H}$, so $\mathfrak{L}_0(\text{ad } \mathfrak{H}) \subset \mathfrak{L}_0(\text{ad } \mathfrak{T}) = \mathfrak{C}_{\mathfrak{L}}(\mathfrak{T})$. Thus, $\mathfrak{L}_0(\text{ad } \mathfrak{H}) = \mathfrak{C}_{\mathfrak{L}}(\mathfrak{T})_0(\text{ad } \mathfrak{H}) = \mathfrak{H}$ and $\mathfrak{H}$ is a Cartan subalgebra of $\mathfrak{L}$ containing $\mathfrak{T}$.

Next let $\mathfrak{H}$ be any Cartan subalgebra of $\mathfrak{L}$. Let $\mathfrak{T}$ be the set of semisimple elements of $\mathfrak{H}$. We claim that $\mathfrak{T}$ is a torus and $\mathfrak{H} = \mathfrak{T}$. Let $x \in \mathfrak{H}$. Then $\text{ad } x|_{\mathfrak{H}}$ is nilpotent, so $\text{ad } x_s|_{\mathfrak{H}} = 0$ and $x_s \in \mathfrak{C}_{\mathfrak{L}}(\mathfrak{H}) \subset \mathfrak{H}$. Thus, $\mathfrak{T} \subset \mathfrak{C}(\mathfrak{H})$ and $x_s \in \mathfrak{T}$ for $x \in \mathfrak{H}$. If x and y are commuting semisimple elements, one sees easily that any linear combination of x and y is semisimple. (Equivalently, two commuting diagonalizable transformations are simultaneously diagonalizable.) Thus, $\mathfrak{T}$ is a subspace of $\mathfrak{H}$, hence a torus, since $xy = 0$ for $x, y \in \mathfrak{T}$. Now let $\mathfrak{N} = \{x \in \mathfrak{H} \mid \text{ad}_{\mathfrak{L}} x \text{ is nilpotent}\} = \{x \in \mathfrak{H} \mid x \text{ is nilpotent}\}$. Then $\mathfrak{N}$ is an ideal of $\mathfrak{H}$, by 3.4.4, and $\mathfrak{H} = \mathfrak{T} \oplus \mathfrak{N}$ since $x \in \mathfrak{H} \Rightarrow x_s \in \mathfrak{T} \subset \mathfrak{H} \Rightarrow x_n \in \mathfrak{N}$ so that $x \in \mathfrak{H} \Rightarrow x = x_s + x_n \in \mathfrak{T} + \mathfrak{N}$. It remains to show that $\mathfrak{N} = \{0\}$, for then $\mathfrak{H} = \mathfrak{T}$. By Engel's theorem, $\text{ad}_{\mathfrak{L}} \mathfrak{N}$

$$K(t, K(x, y)t_a) = K(x, y)K(t, t_a) = a(t)K(x, y).$$

Since $\mathfrak{L}_{-a}$ and $\mathfrak{L}_a$ are dual relative to K (,), it follows from the above that $\mathfrak{L}_{-a}\mathfrak{L}_a = \mathfrak{T}_a$.

We next show that $2a \notin \Sigma$ and $\dim \mathfrak{L}_{\pm a} = 1$. Choose $x \in \mathfrak{L}_{-a}$, $y \in \mathfrak{L}_a$ such that $xy = t_a$. Then $\mathfrak{B} = kx + kt_a + \sum_{i>0} \mathfrak{L}_{ia}$ is stable under ad x and ad y. Thus,

$$0 = \operatorname{Trace}\,[\operatorname{ad} x, \operatorname{ad} y]|_{\mathfrak{B}} = \operatorname{Trace}\,\operatorname{ad} t_a|_{\mathfrak{B}} = -a(t_a) + 0 + \sum_{i>0} d_i ia(t_a) = (a, a)(-1 + \sum_{i>0} d_i i),$$

where $d_i = \dim \mathfrak{L}_{ia}$. Since $(a, a) \neq 0$, by 3.7.2.2, $\sum_{i>0} d_i i = 1$. Thus, dim $\mathfrak{L}_a = 1$ and dim $\mathfrak{L}_{2a} = 0$. Thus, $2a \notin \Sigma$. Replacing a by $-a$, we have dim $\mathfrak{L}_{-a} = 1$.

From the above, it is clear that $\mathfrak{L}^{(a)}$ is a split simple subalgebra of $\mathfrak{L}$ of dimension 3. Assertion 2 is easily verified.

For the following proposition, we note that ad x is nilpotent for $x \in \mathfrak{L}_a$ and $a \neq 0$. For $(\operatorname{ad} x)^i$ maps $\mathfrak{L}_b$ into $\mathfrak{L}_{b+ia}$ for all b, so that $(\operatorname{ad} x)^i = 0$ if $b + ia$ is not a root for $b \in \Sigma \cup \{0\}$. Such an i exists since Σ is finite.

We also recall that if D is a nilpotent derivation of L, then exp D is an automorphism of L (see 2.4.15). In particular, exp ad x is an automorphism of L for ad x nilpotent.

3.7.2.4 Proposition

Let $a \in \Sigma$ and $w_a = (\exp \operatorname{ad} y)(\exp \operatorname{ad} -x)(\exp \operatorname{ad} y)$ where $x \in \mathfrak{L}_{-a}$, $y \in \mathfrak{L}_a$ and $K(x, y) = \dfrac{2}{(a, a)}$. Then w_a is an automorphism of L stabilizing $\mathfrak{T}$, and $w_a|_{\mathfrak{T}}$ is the reflection in $\mathfrak{T}$ relative to K (,) across the hyperplane orthogonal to t_a.

PROOF. Letting $t = \dfrac{2}{(a, a)} t_a$, we have $xy = t$, $xt = -2x$, $yt = 2y$ and $a(t) = 2$. We are to show that $w_a|_{\mathfrak{T}}$ is the reflection across the hyperplane orthogonal to t_a. For this, we show that $tw_a = -t$ and $sw_a = s$ for $s \in \mathfrak{T}$ and $s \perp t_a$, that is, for $a(s) = 0$. Thus, we note that

$$\begin{aligned}
&x \exp \operatorname{ad} y = x + t - y \\
&y \exp \operatorname{ad} (-x) = y + t - x \\
&t \exp \operatorname{ad} y = t - 2y \\
&t \exp \operatorname{ad} (-x) = t - 2x \\
&s \exp \operatorname{ad} y = s \exp \operatorname{ad} (-x) = s \text{ for } a(s) = 0.
\end{aligned}$$

Then

$$\begin{aligned}
&t \exp \operatorname{ad} y \exp \operatorname{ad} (-x) \exp \operatorname{ad} y \\
&= (t - 2y) \exp \operatorname{ad} (-x) \exp \operatorname{ad} y \\
&= (t - 2x - 2(y + t - x)) \exp \operatorname{ad} y \\
&= (t - 2y - 2t) \exp \operatorname{ad} y \\
&= t - 2y - 2y - 2(t - 2y) = -t
\end{aligned}$$

and

$$\begin{aligned}
&s \exp \operatorname{ad} y \exp \operatorname{ad} (-x) \exp \operatorname{ad} y \\
&= s \exp \operatorname{ad} (-x) \exp \operatorname{ad} y \\
&= s \exp \operatorname{ad} y = s.
\end{aligned}$$

3.7.2.5 Definition

The group W of automorphisms generated by $\{w_a \mid a \in \Sigma\}$ is called a *Weyl group* of L with respect to $\mathfrak{T}$.

It is of no grave consequence for us that W may depend superficially on the particular choice of the x, y used in defining the W_a.

For $w \in W$, let w' denote the inverse of the adjoint of $w|_{\mathfrak{T}}$, that is, $(aw')(t) = a(tw^{-1})$ for $a \in \mathfrak{T}^*$. Then $(\mathfrak{L}_b)w = \mathfrak{L}_{bw'}$ for $b \in \Sigma$, so that $\Sigma w' \subset \Sigma$. Thus, $\mathfrak{T}_{\mathrm{Q}}^* w' \subset \mathfrak{T}_{\mathrm{Q}}^*$. We let $w^* = w'|_{\mathfrak{T}_{\mathrm{Q}}^*}$. Since the reflections $w_a|_{\mathfrak{T}}$ $(a \in \Sigma)$ preserve $\mathrm{K}(\ ,\ )$, so do the $w|_{\mathfrak{T}}$ $(w \in W)$. Thus, we have $t_b w = t_{bw^*}$ for $b \in \mathfrak{T}_{\mathrm{Q}}^*$ and $w \in \mathfrak{W}$. It follows that the following diagram is commutative for $w \in \mathfrak{W}$:

$$\begin{array}{ccc} \mathfrak{T}_Q & \xrightarrow{\;w\;} & \mathfrak{T}_Q \\ \downarrow & & \downarrow \\ \mathfrak{T}_Q^* & \xrightarrow[\;w^*\;]{} & \mathfrak{T}_Q^* \end{array}$$

Figure 4.

For $a \in \Sigma$, w_a^* is therefore the reflection at a in $\mathfrak{T}_Q^*$ with respect to $(\, ,)$, since $w_a|_{\mathfrak{T}}$ is the reflection at t_a in $\mathfrak{T}$ with respect to $K(\, ,)$.

3.7.2.6 Definition

The group W^* of automorphisms of $\mathfrak{T}_Q^*$ generated by the reflections $\{w_a^* \mid a \in \Sigma\}$ described above is called the *Weyl group* of $(\mathfrak{T}_Q^*, \Sigma)$.

We obviously have the following proposition.

3.7.2.7 Proposition

The mapping $w \mapsto w^*$ is a homomorphism from W onto W^* and $\mathfrak{L}_b w = \mathfrak{L}_{bw^*}$ for $w \in W$, $b \in \Sigma$.

3.7.2.8 Proposition

Let $a, b \in \Sigma$ and $(a, b) > 0$. Then $a - b \in \Sigma \cup \{0\}$. If $a \neq b$, then $\mathfrak{L}_a \mathfrak{L}_{-b} = \mathfrak{L}_{a-b}$.

PROOF. Suppose $a - b \notin \Sigma \cup \{0\}$ or that $a \neq b$ and $\mathfrak{L}_a \mathfrak{L}_{-b} = \{0\}$. Then $\mathfrak{B} = \sum_{i=0}^{\infty} \mathfrak{L}_{a+ib}$ is stable under ad $\mathfrak{L}_{\pm b}$. Since ad $t_b \in [\text{ad}\, \mathfrak{L}_{-b}, \text{ad}\, \mathfrak{L}_b]$, it follows as usual that $0 = \text{Trace ad}\, t_b|_{\mathfrak{B}} = \sum d_i(a + b_i)(t_b) = \sum d_i(a, b) + (\sum i d_i)(b, b)$ where $d_i = \dim \mathfrak{L}_{a+ib}$. Since $(b, b) > 0$, we must have $(a, b) \leq 0$, a contradiction. Thus, $a - b \in \Sigma \cup \{0\}$, and $\mathfrak{L}_a \mathfrak{L}_{-b} \neq 0$ if $a \neq$ b. Since $\dim \mathfrak{L}_{a-b} \leq 1$ for $a \neq b$, $\mathfrak{L}_a \mathfrak{L}_{-b} = \mathfrak{L}_{a-b}$ for $a \neq b$.

We now summarize some properties of the space $\mathfrak{T}_Q^*$, the form $(\, ,)|_{\mathfrak{T}_Q^*}$, and the subset Σ of $\mathfrak{T}_Q^*$.

3.7.2.9 Theorem

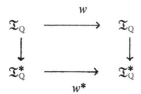

$\mathfrak{T}_Q^*$ is a finite-dimensional vector space over $\mathbb{Q}$ and $(\, ,)|_{\mathfrak{T}_Q^*}$ is a $\mathbb{Q}$-valued positive-definite symmetric bilinear form on $\mathfrak{T}^*_{\mathbb{Q}}$. The set Σ is a finite subset of $\mathfrak{T}^*_{\mathbb{Q}}$ satisfying the following properties:

1. Σ spans $\mathfrak{T}^*_{\mathbb{Q}}$ over $\mathbb{Q}$;
2. $0 \notin \Sigma$;
3. for $a \in \Sigma$, the reflection w^*_a of $\mathfrak{T}^*_{\mathbb{Q}}$ at a maps Σ into Σ;
4. $(a, b) < 0 \Rightarrow a + b \in \Sigma \cup \{0\}$;
5. $a \in \Sigma \Rightarrow 2a \notin \Sigma$.

3.7.2.10 Definition

The pair $(\mathfrak{T}^*_{\mathbb{Q}}, \Sigma)$ together with the form $(\, , \,)|_{\mathfrak{T}^*_{\mathbb{Q}}}$ is called the *root system of $\mathfrak{L}$ relative to $\mathfrak{T}$.*

Root systems are classified in subsection 3.7.3, independently of the theory of Lie algebras. In 3.7.4, it is shown that the Lie algebra $\mathfrak{L}$ is determined up to isomorphism by its root system, so that the isomorphism classes of split semisimple Lie algebras over a field k of characteristic 0 correspond bijectively to the isomorphism classes of root systems. The isomorphism classes are described in 3.7.5.

3.7.3 Abstract Root Systems

3.7.3.1 Definition of Root Systems

In 3.7.2, we introduced the root system of a split semisimple Lie algebra of characteristic 0. This is a root system in the sense of this subsection. The objective here is to describe and classify root systems independently of the theory of Lie algebras. The axioms for a root system given here are, in view of 3.7.3.3.1, essentially equivalent to those given in [7].

3.7.3.1.1 Definition

A *root system* is a pair (V, Σ) where V is a finite-dimensional vector space over $\mathbb{Q}$ together with a positive-definite symmetric bilinear form $(\, , \,)$, and where Σ is a finite subset of V such that

1. Σ spans V;
2. $0 \notin \Sigma$;
3. Σ is stable under r_a for $a \in \Sigma$, where r_a is the reflection in V across the hyperplane orthogonal to a;

4. $(a, b) < 0 \Rightarrow a + b \in \Sigma \cup \{0\}$ for $a, b \in \Sigma$;
5. $a \in \Sigma \Rightarrow 2a \notin \Sigma$.

If (V, Σ) satisfies all of the above conditions except 5, it is a *weak root system.*

3.7.3.1.2 Definition

If (V, Σ), (V', Σ') are weak root systems, an *isomorphism* from (V, Σ) to (V', Σ') is an isomorphism $f: V \to V'$ of vector spaces such that $f(\Sigma) = \Sigma'$. If $(V, \Sigma) = (V', \Sigma')$, such an f is an *automorphism* of (V, Σ). The set of automorphisms of (V, Σ) is denoted Aut (V, Σ) or Aut Σ.

3.7.3.1.3 Definition

Let (V, Σ) be a weak root system. The *rank* of (V, Σ) is dim V. The elements of Σ are *roots.*

We assume in subsections 3.7.3.2 through 3.7.3.8 that (V, Σ) is a root system. Weak root systems are briefly discussed in 3.7.3.9.

3.7.3.1.4 Proposition

Let $f: (V, \Sigma) \to (V', \Sigma')$ be an isomorphism of root systems. Then $fr_{af} = r_a f$ for $a \in \Sigma$.

PROOF. Let $s = r_{af}$ and define $t: V' \to V'$ by $bft = br_a f$ for $b \in V$. We claim that $s = t$. Since s and t stabilize the finite set Σ' of generators for V', the group G generated by s and t is finite. Now G stabilizes the one-dimensional subspace $W' = \mathbb{Q}(af)$ of V'. It follows that G stabilizes some complementary subspace W'' of V', so that $V' = W' \oplus W''$. For this, we could invoke Maschke's theorem (see [9]), or take a linear mapping $\pi: V' \to W' \simeq \mathbb{Q}$ such that $\pi|_{W'} \neq 0$ and let W'' be the kernel of the nonzero mapping $x \mapsto \sum_{g \in G} ((xg^{-1})\pi)g$ from V' onto $W' \simeq \mathbb{Q}$. Now $s|_{W'} = t|_{W'} = -\mathrm{id}_{W'}$ and, since s and t are defined by reflections, $s|_{W''} = \mathrm{id}_{W''} = t|_{W''}$. Thus, $s = t$ and $bfr_{af} = br_a f$ for $b \in V$, $a \in \Sigma$.

3.7.3.1.5 Corollary

$r_{aw} = w^{-1} r_a w$ for $w \in$ Aut (V, Σ).

3.7.3.1.6 Proposition
The only scalar multiples of a root b which are roots are $\pm b$.

PROOF. Suppose not and let b be a root such that $\mathbb{Q}b \cap \Sigma$ contains more than two elements. We may assume that $(a, a) \geq (b, b)$ for $a \in \mathbb{Q}b \cap \Sigma$. Choose $a \in \mathbb{Q}b \cap (\Sigma - \{b, -b\})$ with (a, a) minimal. Write $a = cb$ with $a \in \mathbb{Q}$. We may assume that $c > 0$, replacing b by $-b$ if necessary. Now $(a, b) > 0$ and $a - b \in \Sigma$ by 3.7.3.1.1. Thus, $(c - 1)b \in \Sigma$ and $(c - 1)b = b$ by the minimality of (a, a). Thus, $c = 2$. But then b and $2b$ are roots, in contradiction with 3.7.3.1.1.

3.7.3.2 Half-Systems and Simple Systems of Σ

3.7.3.2.1 Definition
An element t of the dual space V^* of V is *regular* if $t(a) \neq 0$ for $a \in \Sigma$. A *half-system* of Σ is a subset Σ^+ of the form $\Sigma^+ = \Sigma^+(t)$ where t is a regular element of V^* and $\Sigma^+(t) = \{a \in \Sigma \mid t(a) > 0\}$. An element a of a half-system Σ^+ is *simple* in Σ^+ if a cannot be expressed as $a = b + c$ with $b, c \in \Sigma^+$. The set of simple roots of a half-system Σ^+ is denoted $\pi(\Sigma^+)$. A *simple system* of Σ is a subset π of Σ of the form $\pi = \pi(\Sigma^+)$ for some half-system Σ^+ of Σ.

Since $a \in \Sigma$ iff $-a \in \Sigma$, a subset Σ^+ of Σ is a half-system of Σ iff $\Sigma^- = \{-a \mid a \in \Sigma^+\}$ is a half-system of Σ. In fact, $\Sigma^+ = \Sigma^+(t)$ iff $\Sigma^- = \Sigma^+(-t)$, and Σ is the disjoint union of Σ^+ and Σ^-.

3.7.3.2.2 Proposition
Let π be a simple system of Σ. Then π is a basis for V and $(a, b) \leq 0$ for any two distinct $a, b \in \pi$. Each element b of Σ may be expressed as $b = \pm\sum_{a\in\pi} c_a a$ where the c_a are nonnegative integers $(a \in \pi)$. The set $\Sigma^+ = \{b \in \Sigma \mid b = \sum_{a\in\pi} c_a a$ with $c_a \geq 0$ for $a \in \pi\}$ is the unique half-system of Σ containing π.

PROOF. Let π be the set of simple roots of a half-system $\Sigma^+ = \Sigma^+(t)$, t a regular element of V^*. If $a \in \Sigma^+$ and a is not simple, then $a = b + c$ with $b, c \in \Sigma^+$. Thus, $t(a) > t(b), t(c)$. Since Σ^+ is finite, this observation can be used in an induction argument to show that $\Sigma^+ = \{b \in \Sigma \mid b =$

$\sum_{a\in\pi} c_a a$ with c_a a nonnegative integer for $a \in \pi\}$. Since $\Sigma = \Sigma^- \cup \Sigma^+$, any element b of Σ can be expressed as $b = \pm\sum_{a\in\pi} c_a a$ where the c_a are nonnegative integers.

It remains to show that π is a basis for V such that $(a, b) \leq 0$ for distinct $a, b \in \pi$. By the first paragraph, π spans V. For $a, b \in \pi$ and $(a, b) > 0$, we see that $a = b$ as follows. The vectors $a - b$ and $b - a$ are in $\Sigma \cup \{0\}$ by 3.7.3.1.1. But a and b are simple and $a = (a - b) + b$, $b = a + (b - a)$. Thus $b - a \notin \Sigma^+$ and $a - b \notin \Sigma^+$. Thus, $a = b$. Thus, $(a, b) \leq 0$ if $a \neq b$, $a, b \in \pi$. We now show that π is a linearly independent set. Thus, let $\sum_{a\in\pi} c_a a = 0$ with $c_a \in \mathbf{Q}$ for $a \in \pi$. Let $b = \sum_{c_a\geq 0} c_a a$ and $c = \sum_{c_a\leq 0} c_a a$. Then $b + c = 0$ and $0 = (b + c,\ b + c) = (b,\ b) + 2(b,\ c) + (c,\ c)$. Since $(b, c) = \sum c_a c_{a'}(a, a')$ with $c_a \geq 0$, $c_{a'} \leq 0$ and $(a, a') \leq 0$, we have $(b, c) \geq 0$ and, by the above equation, $(b, b) = (c, c) = 0$. Thus, $b = c = 0$. Now $t(a) > 0$ for $a \in \pi$. This together with the equations

$$0 = t(b) = \sum_{c_a\geq 0} c_a t(a),$$

$$0 = t(c) = \sum_{c_a\leq 0} c_a t(a),$$

imply that $c_a = 0$ for $a \in \pi$. Thus, π is linearly independent.

3.7.3.2.3 Proposition
Let $a \in \Sigma$. Then $a \in \pi$ for some simple system π.

PROOF. It suffices to find a regular element t of V^* such that $t(a) > 0$ and $t(a) \leq t(b)$ for $b \in \Sigma^+(t)$. For this, let $e_1, \ldots, e_r$ be a basis for V with $a = e_1$. Let W be the subspace spanned by $e_2, \ldots, e_r$. Let $T: V \to W$ be the projection such that $e_1 T = 0$, $e_i T = e_i$ for $i \geq 2$. Let $\Sigma_W = \{bT \mid b \in \Sigma - \mathbf{Q}a\}$ and note that $0 \notin \Sigma_W$. Choose $t_W \in W^*$ such that t_W does not vanish at any element of the finite set Σ_W. This is possible since any finite set $\{w_1, \ldots, w_n\}$ of nonzero elements of W defines a nonzero polynomial function $w: W^* \to \mathbf{Q}$, given by $w(t) = t(w_1) \ldots t(w_n)$ for $t \in W^*$, whose nonzeros t do not vanish at any w_i, $1 \leq i \leq n$. Define $t_0 \in V^*$ by $t_0(e_1) = 0$, $t_0|_W = t_W$. Then $t_0(b) = t_W(bT) \neq 0$ for $b \in \Sigma - \mathbf{Q}a$, and

$t_0(a) = 0$. For any positive rational d, define t_d by $t_d(e_1) = t_d(a) = d$, $t_d(e_i) = 0$ for $2 \leq i \leq n$. Let $t = t_0 + t_d$ where $d > 0$ is chosen such that $t(a) = d < |t(b)|$ for $b \in \Sigma - \mathbf{Q}a$. Then $t(a) > 0$, and $t(a) \leq t(b)$ for $b \in \Sigma^+(t)$. For if $b \in \Sigma^+ - \mathbf{Q}a$, then $t(a) < t(b)$; and if $b \in \mathbf{Q}a$, then $b = \pm a$.

3.7.3.3 Pairs of Roots

For $a \in V - \{0\}$, the reflection r_a about the hyperplane orthogonal to a is given by $vr_a = v - a^*(v)a$ where $a^*(v) = 2\dfrac{(a, v)}{(a, a)}$.

3.7.3.3.1 Proposition

For $a, b \in \Sigma$, $a^*(b)$ is an integer.

PROOF. By 3.7.3.2.3, $a \in \pi$ for some simple system π of Σ. Suppose that $b \in \pi$. If $b = a$, then $a^*(b) = 2$. If $b \neq a$, the root $br_a = b - a^*(b)a$ has a unique decomposition $b - a^*(b)a = \pm\sum_{d \in \pi} c_d d$ where the c_d are non-negative integers. Thus, $a^*(b) = \mp c_a \in \mathbf{Z}$. We have therefore shown that $a^*(b) \in \mathbf{Z}$ for $b \in \pi$. But a^* is linear and every element b of Σ is an integral linear combination of elements of π. Thus, $a^*(b) \in \mathbf{Z}$ for $b \in \Sigma$.

We now consider a pair a, b of linearly independent roots with $(a, b) \leq 0$ and $|a| \geq |b|$. Since $a^*(b)b^*(a) = 4\dfrac{(a, b)^2}{|a|^2|b|^2} = 4\cos^2\theta < 4$ where θ is the angle between a, b, there are only four cases:

case	$a^*(b)$	$b^*(a)$	$\dfrac{(a, a)}{(b, b)}$	$\cos\theta$	$\lvert\theta\rvert$
0	0	0		0	90°
1	-1	-1	1	$-\dfrac{1}{2}$	120°
2	-1	-2	2	$-\dfrac{\sqrt{2}}{2}$	135°
3	-1	-3	3	$-\dfrac{\sqrt{3}}{2}$	150°

3.7.3.3.2 Proposition
Let $a, b \in \Sigma$ with $(a, b) < 0$. Then $(a, a) + (b, b) + 4(a, b) \leq 0$.

PROOF. If $a = -b$, this is trivial. Suppose $a \neq -b$. Then a and b are linearly independent. We may assume that $|a| \geq |b|$. From the above table, we have

$$\frac{(a, a)}{(b, b)} + 2\frac{(a, b)}{(b, b)} = 0,$$

$$\frac{(b, b)}{(b, b)} + 2\frac{(a, b)}{(b, b)} \leq 0.$$

Adding these together, we have

$$\frac{1}{(b, b)}((a, a) + (b, b) + 4(a, b)) \leq 0,$$

and the assertion follows.

3.7.3.4 The Weyl Group

3.7.3.4.1 Definition
For any subset S of Σ, we let W_S be the group generated by the set $\{r_a \mid a \in S\}$ of reflections of V. The group $W_\Sigma = W$ is called the *Weyl group* of Σ.

Obviously, W is a subgroup of $\operatorname{Aut}(V, \Sigma)$ and the elements of W preserve $(\, , \,)$.

If Σ^+ is a half-system of Σ and π the corresponding simple system, then $\Sigma^+ w$ is a half-system and πw the corresponding simple system for $w \in W$.

3.7.3.4.2 Proposition
W acts transitively on the set of half-systems and on the set of simple systems of Σ. If π is a simple system of Σ, then $W_\Sigma = W_\pi$.

PROOF. Let Σ^+ be a half-system of Σ. Let Σ_1^+ be another half-system of Σ, and let π_1 be the corresponding simple system. Choose $w \in W_\pi$ such that $\Sigma_1^+ w \cap \Sigma^+$ has maximal cardinality. If $\Sigma_1^+ w \neq \Sigma^+$, there exists an $a \in \pi - \Sigma_1^+ w \cap \Sigma^+$. Then $\Sigma^+ - \{a\}$ is r_a-stable. For let $d \in \Sigma^+ - \{a\}$. Then $d = \sum_{b \in \pi} c_b b$ with $c_b > 0$ for some $b \neq a$, and $dr_a = d - a^*(d)a = \sum_{b \in \pi} c_b' b$ with $c_b' = c_b$ for $b \neq a$. Thus, $c_b' > 0$ for some $b \in \pi$, so $dr_a \in \Sigma^+ - \{a\}$, by 3.7.3.2.2. Since $\Sigma^+ - \{a\}$ is r_a-stable, $\Sigma_1^+ wr_a \cap \Sigma^+$ contains $(\Sigma_1^+ w \cap \Sigma^+) \cup \{a\}$, a contradiction in view of the maximality of the cardinality of $\Sigma_1^+ w \cap \Sigma^+$. Thus, $\Sigma_1^+ w = \Sigma^+$ and W_π acts transitively on the set of half-systems of Σ. Consequently, W_π acts transitively on the set of simple systems of Σ.

It remains to show that $W_\Sigma = W_\pi$. Thus, let $a \in \Sigma$. Then $a \in \pi_1$ for some simple system π_1 of Σ. Choose $w \in W_\pi$ such that $\pi_1 w = \pi$. Then $aw = b \in \pi$ and $w^{-1} r_a w = r_b$, by 3.7.3.1.5. Thus, $r_a \in wW_\pi w^{-1} = W_\pi$. Thus, $W_\Sigma = W_\pi$.

One can show that W_Σ acts simply transitively on the set of half-systems of Σ. We do not need this, and refer the reader to [7, p. 242] for a proof.

3.7.3.4.3 Theorem

Let (V, Σ), (V', Σ') be root systems, π a simple system of Σ, and π' a simple system of Σ'. Then any bijection f from π to π' such that $(af)^*(bf) = a^*(b)$ for all $a, b \in \pi$ can be extended to an isomorphism f from (V, Σ) to (V', Σ'). Conversely, if f is an isomorphism from (V, Σ) to (V', Σ'), then $(af)^*(bf) = a^*(b)$ for all $a, b \in \Sigma$.

PROOF. First, let f be an isomorphism from (V, Σ) to (V', Σ') and let a, $b \in \Sigma$. We may assume that $a \neq \pm b$, so that a and b are linearly independent. Now $br_a f = bfr_{af}$, by 3.7.3.1.4. Since $br_a = b - a^*(b)a$ and $(bf)r_{af} = bf - (af)^*(bf)af$, and since af, bf are linearly independent, $(af)^*(bf) = a^*(b)$.

Next, let f be a bijection from π to π' such that $(af)^*(bf) = a^*(b)$ for $a, b \in \pi$. Extending f to a vector space isomorphism from V to V' and reasoning as in the above paragraph, we have $(bf)r_{af} = (br_a)f$ for a,

$b \in \pi$. Since π is a basis for V, the following diagram is commutative for $a \in \pi$:

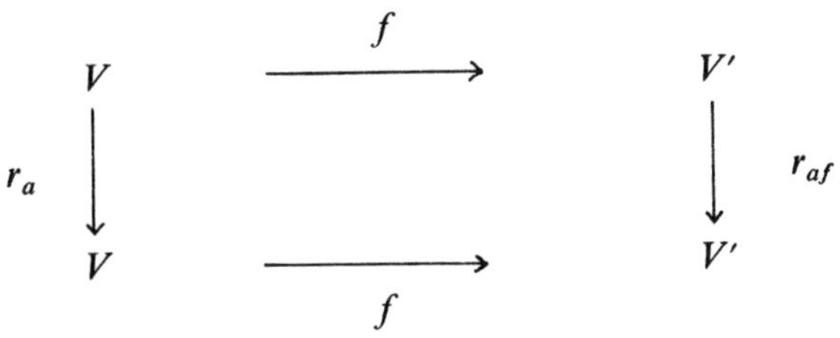

Figure 5.

Now $\pi f = \pi'$, $\{r_a \mid a \in \pi\}$ generates W_Σ and $\{r_{a'} \mid a' \in \pi'\}$ generates $W_{\Sigma'}$. By the commutativity of the above diagram, the following diagram is commutative where $w = r_{a_1} \ldots r_{a_n}$ $(a_i \in \pi)$ is any expression of an element of W_Σ and $w' = r_{a_1 f} \ldots r_{a_n f}$ the corresponding expression of an element of $W_{\Sigma'}$.

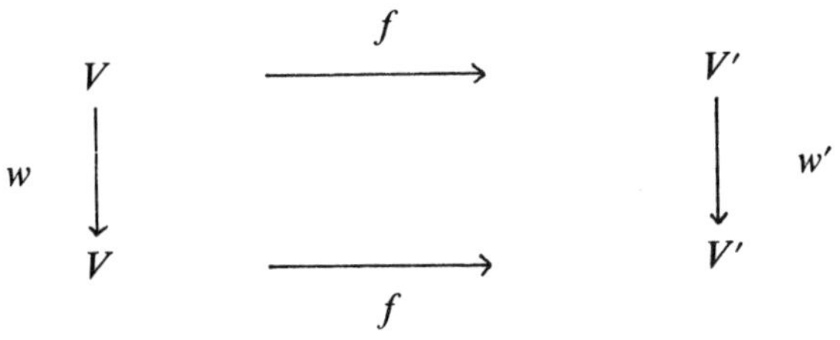

Figure 6.

Now every element of W_Σ has such an expression w and every element of $W_{\Sigma'}$ has such an expression w'. Since $\pi f = \pi'$, $\Sigma = \{bw \mid b \in \pi$ and $w \in W_\Sigma\}$ and $\Sigma' = \{b'w' \mid b' \in \pi'$ and $w' \in W_{\Sigma'}\}$, by 3.7.3.2.3 and 3.7.3.4.2, it follows that $\Sigma f = \Sigma'$, hence that f is an isomorphism from (V, Σ) to (V', Σ').

3.7.3.5 Root Sums

3.7.3.5.1 Definition

Two elements a, b of Σ are *adjacent* if they are distinct and $(a, b) \neq 0$.

For $S \subset \Sigma$, a *chain* in S is a sequence $a_0, \ldots, a_m$ of elements of S such that a_i, a_{i+1} are adjacent for $0 \leq i \leq m-1$. A subset S of Σ is *connected* if any two distinct elements $a, b \in S$ can be joined by a chain $a = a_0, \ldots, a_m = b$. In a connected set S we have a metric $d_S(a, b) = \min\{m \mid a, b$ can be joined by a chain $a = a_0, \ldots, a_m = b\}$ for $a, b \in S$.

3.7.3.5.2 Proposition
Let S be connected and nonempty. Then $S - \{b\}$ is connected for some $b \in S$.

PROOF. We may assume that S has at least two elements. Choose $a, b \in S$ with $d_S(a, b)$ maximal. We claim that $S - \{b\}$ is connected. Thus, let $c \in S - \{b\}$ and let $a = a_0, \ldots, a_m = c$ be a chain in S with $m = d_S(a, c)$. Then $m \leq d_S(a, b)$, so that no a_i can be b. Thus, a and c can be joined by a chain in $S - \{b\}$ for any $c \in S - \{b\}$. Thus, $S - \{b\}$ is connected.

3.7.3.5.3 Definition
The *extreme roots* of a connected set S are those $b \in S$ such that $S - \{b\}$ is connected.

3.7.3.5.4 Definition
A *standard subset* of Σ is a subset S of Σ such that S is contained in some half-system Σ^+ of Σ and $(a, b) \leq 0$ for distinct $a, b \in S$.

3.7.3.5.5 Example
A simple system π of Σ is a standard subset of Σ, by 3.7.3.2.2.

The following proposition is proved in precisely the way that we proved that a simple system π of Σ is a linearly independent set.

3.7.3.5.6 Proposition
A standard subset S of Σ is a linearly independent set.

3.7.3.5.7 Proposition
Let S be a connected standard subset of Σ. Let $[S] = \sum_{a \in S} a$. Then $[S]$ is a root. If the elements of S are of a fixed length d, then $[S]$ is of length d also.

PROOF. Suppose not, and let S be a counterexample having as few elements (at least two, obviously) as possible. Let b be an extreme root of S. Since $S - \{b\} = T$ is connected, $[T]$ is a root by the minimality of S. But $([T], b) < 0$ by the hypothesis, so $[S] = [T] + b$ is a root or 0 by 3.7.3.1.1. But S is linearly independent, so $[S]$ is not 0. Suppose now that the elements of S are all of length d. By the minimality of S, $|[T]| = |b| = d$. Since $([T], b) < 0$, $|[T]| = |b|$ implies that

$$2\frac{([T], b)}{([T], [T])} = 2\frac{([T], b)}{(b, b)} = -1,$$

by 3.7.3.3. Thus, $([T] + b, [T] + b) = ([T], [T]) = (b, b)$. Now $|[S]| = |[T] + b| = |b| = d$.

For the remainder of the section, we fix a simple system π of positive roots of Σ. For $b \in \Sigma$, let $b = \pm\sum_{a \in \pi} c_a a$ where the c_a are nonnegative integers.

3.7.3.5.8 Definition

The *level* of b with respect to π is $L(b) = \sum_{a \in \pi} c_a$. The *support* of b with respect to π is $\pi(b) = \{a \in \pi \mid c_a \neq 0\}$.

3.7.3.5.9 Lemma

For $b \in \Sigma^+ - \pi$, there exists an $a \in \pi(b)$ such that $b - a \in \Sigma^+$. For such an a, $L(b) = L(b - a) + 1$.

PROOF. Since $(b, b) > 0$ and $b = \sum_{a \in \pi} c_a a$, the c_a being nonnegative, $(b, a) > 0$ for some $a \in \pi$. For such an a, $b - a$ is a root, by 3.7.3.1.1. Since some coefficient in the expression of $b - a$ as a linear combination of elements of π is positive, $b - a \in \Sigma^+$. That $L(b) = L(b - a) + 1$ is trivial.

3.7.3.5.10 Proposition

The support $\pi(b)$ of b is connected.

PROOF. We may take $b \in \Sigma^+$. Suppose that the assertion is false and let $b \in \Sigma^+$ be a counterexample with $b = \sum_{a \in \pi} c_a a$ and $L(b) = \sum_{a \in \pi} c_a$ minimal. Since $\pi(b)$ is not connected, $\pi(b) = \pi_1 \cup \pi_2$ where π_1, π_2 are nonempty and $(a_1, a_2) = 0$ for $a_1 \in \pi_1$, $a_2 \in \pi_2$. Since $(b, b) > 0$ and the c_a are nonnegative, there exists an $a_1 \in \pi(b)$ such that $(b, a_1) > 0$, so that $b - a_1 \in \Sigma^+$. We may assume that $a_1 \in \pi_1$. Since $L(b - a_1) < L(b)$, the support $\pi(b - a_1)$ of $b - a_1$ is connected. This set must be $(\pi_1 - \{a_1\}) \cup \pi_2$ and it follows that $\pi_1 = \{a_1\}$ and $c_{a_1} = 1$. Now

$$br_{a_1} = a_1 r_{a_1} + (\sum_{a \in \pi_2} c_a a) r_{a_1} = -a_1 + \sum_{a \in \pi_2} c_a a,$$

since $(a_1, a) = 0$ for $a \in \pi_2$. But this contradicts 3.7.3.2.2.

3.7.3.6 Subsystems

3.7.3.6.1 Definition

A *subsystem* of (V, Σ) is a pair (V', Σ') together with $(\ ,\ )|_{V'}$ where V' is a subspace of V, $\Sigma' = \Sigma \cap V'$, and V' is spanned by Σ'. Since a subspace V' is r_a-stable for $a \in V' - \{0\}$, a subsystem (V', Σ') of (V, Σ) is a root system. Two subsystems (V', Σ'), (V'', Σ'') are *complementary* if $V = V' \oplus V''$ and $\Sigma = \Sigma' \cup \Sigma''$. A subsystem (V', Σ') is a *direct factor* if there exists a subsystem complementary to it. The root system (V, Σ) is *irreducible* if it has no proper direct factor.

3.7.3.6.2 Proposition

Let (V', Σ'), (V'', Σ'') be complementary subsystems. Then $V' \perp V''$.

PROOF. Since $W_{\Sigma'}$ stabilizes Σ', it stabilizes $\Sigma'' = \Sigma - \Sigma'$, thus also V''. Now

$$v'' r_{a'} = v'' - 2\frac{(a', v'')}{(a', a')} a' \in V''$$

for $v'' \in V''$ and $a' \in \Sigma' \subset V'$. Thus, $(a', v'') = 0$ for $a' \in \Sigma'$ and $v'' \in V''$, so that $V' \perp V''$.

3.7.3.6.3 Corollary
If some simple system π of Σ is connected, then (V, Σ) is irreducible.

PROOF. Otherwise (V, Σ) has proper complementary subsystems (V', Σ'), (V'', Σ''). Let π', π'' be simple systems of Σ', Σ'' respectively. Then $\pi' \cup \pi''$ is a nonconnected simple system of Σ. But there exists $w \in W_\Sigma$ such that $\pi' \cup \pi'' = \pi w$, a contradiction since πw is connected.

3.7.3.6.4 Proposition
Let (V_1, Σ_1), (V_2, Σ_2) be direct factors of (V, Σ). Then $(V_1 \cap V_2, \Sigma_1 \cap \Sigma_2)$ and $(V_1 + V_2, \Sigma_1 \cup \Sigma_2)$ are direct factors of (V, Σ).

PROOF. Let (V_i', Σ_i') be a subsystem of (V, Σ) complementary to (V_i, Σ_i) $(i = 1, 2)$. We claim that $(V_1 \cap V_2, \Sigma_1 \cap \Sigma_2)$ and $(V_1' + V_2', \Sigma_1' \cup \Sigma_2')$ are complementary subsystems of (V, Σ). Since $\Sigma = \Sigma_i \cup \Sigma_i'$ $(i = 1, 2)$, $\Sigma = (\Sigma_1 \cap \Sigma_2) \cup (\Sigma_1' \cup \Sigma_2')$. Thus, $V = (V_1 \cap V_2) + (V_1' + V_2')$. It remains to show that the sum is direct. Clearly we have $\dim V \leq \dim (V_1 \cap V_2) - \dim (V_1' \cap V_2') + \dim V_1' + \dim V_2'$, and it remains to show equality. But, by the symmetry, we have also $\dim V \leq \dim (V_1' \cap V_2') - \dim (V_1 \cap V_2) + \dim V_1 + \dim V_2$. Since $\dim V_i = \dim V - \dim V_i'$, it follows that $\dim V \leq \dim (V_1' \cap V_2') - \dim (V_1 \cap V_2) - \dim V_1' - \dim V_2' + 2 \dim V$, or that $\dim (V_1 \cap V_2) - \dim (V_1' \cap V_2') + \dim V_1' + \dim V_2' \leq \dim V$. Thus, we have equality.

3.7.3.6.5 Definition
For $S \subset \Sigma$, we let V_S be the subspace of V spanned by S and $\Sigma_S = \Sigma \cap V_S$.
Obviously, (V_S, Σ_S) is a subsystem of (V, Σ).

3.7.3.6.6 Theorem
Let π be the set of simple roots of a half-system Σ^+ of Σ. Let $\pi_1, \ldots, \pi_n$ be the distinct connected components of π. Then the irreducible direct factors of (V, Σ) are the subsystems $(V_{\pi_i}, \Sigma_{\pi_i})$ $(i = 1, \ldots, n)$, $V = V_{\pi_1} \oplus \ldots \oplus V_{\pi_n}$ (orthogonal direct sum) and $\Sigma = \Sigma_{\pi_1} \cup \ldots \cup \Sigma_{\pi_n}$.

PROOF. $V = V_{\pi_1} \oplus \ldots \oplus V_{\pi_n}$ (orthogonal direct sum) since π is a basis for V. That $\Sigma = \Sigma_{\pi_1} \cup \ldots \cup \Sigma_{\pi_n}$ follows from 3.7.3.5.10. Thus, the $(V_{\pi_i}, \Sigma_{\pi_i})$ are direct factors. They are irreducible by 3.7.3.6.3. Now let (V', Σ') be any irreducible direct factor. Then the subsystems $(V' \cap V_{\pi_i}, \Sigma' \cap \Sigma_{\pi_i})$ are direct factors, and (V', Σ') is the direct sum of the $(V' \cap V_{\pi_i}, \Sigma' \cap \Sigma_{\pi_i})$ since $\Sigma' = \bigcup \Sigma' \cap \Sigma_{\pi_i}$. Since (V', Σ') is irreducible, it follows that $(V', \Sigma') = (V_{\pi_i}, \Sigma_{\pi_i})$ for some i.

3.7.3.6.7 Corollary

Let $a \in \Sigma_{\pi_i}$, $b \in \Sigma_{\pi_j}$ with $i \neq j$, notation as in 3.7.3.6.6. Then $a + b \notin \Sigma$.

3.7.3.6.8 Corollary

(V, Σ) is irreducible iff a simple system π of Σ is connected.

3.7.3.7 Classification of Standard Subsets of Σ

In 3.7.3.5 we introduced the notion of standard subset of Σ and gave as an example any simple system π of Σ. We now classify the standard sets by attaching to each a diagram and showing that the diagrams fall into certain types. The ensuing classification of the possible simple systems leads to the classification of root systems given in 3.7.3.8.

We begin with some elementary properties of standard subsets of Σ.

3.7.3.7.1 Definition

A *cycle* in Σ is a standard subset $\{a_1, \ldots, a_n\}$ of Σ such that $n > 3$ and $(a_1, a_2) < 0, \ldots, (a_{n-1}, a_n) < 0, (a_n, a_1) < 0$.

3.7.3.7.2 Proposition

There is no cycle in Σ.

PROOF. Suppose that $\{a_1, \ldots, a_n\}$ is a cycle and let $a' = a_3 + \ldots + a_n$. Then $a' \in \Sigma$, by 3.7.3.5.7, and $\{a_1, a_2, a'\}$ is a cycle. Thus, we may assume that $n = 3$. Let $a = a_1 + a_2 + a_3$. Then

$$\begin{aligned} 2(a, a) = {} & (a_1, a_1) + (a_2, a_2) + 4(a_1, a_2) \\ & + (a_2, a_2) + (a_3, a_3) + 4(a_2, a_3) \\ & + (a_3, a_3) + (a_1, a_1) + 4(a_3, a_1). \end{aligned}$$

But then $2(a, a) \leq 0$, by 3.7.3.3.2, a contradiction.

3.7.3.7.3 Proposition

Let S be a standard set, $b \in S$. Then b is adjacent to at most three elements of S.

PROOF. Suppose that b is adjacent to a_1, a_2, a_3, a_4:

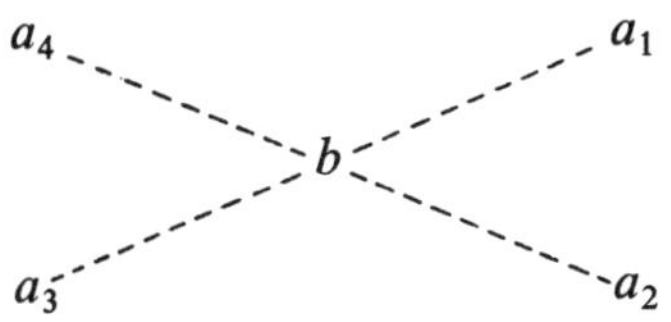

Figure 7.

Since there are no cycles in S, $(a_i, a_j) = 0$ for $i \neq j$. Letting $a = 2b + \sum_1^4 a_i$, we have

$$(a, a) = 4(b, b) + \sum_1^4 (a_i, a_i) + \sum_1^4 4(b, a_i)$$
$$= \sum_1^4 ((b, b) + (a_i, a_i) + 4(b, a_i)) \leq 0,$$

by 3.7.3.3.2, a contradiction.

3.7.3.7.4 Definition

Let S be a standard subset, and let S' be a connected subset of S. Then $S/S' = (S - S') \cup \{[S']\}$ where we recall that $[S']$ is the root $\sum_{a \in S'} a$. The *canonical* mapping from S to S/S' is defined by $s \mapsto s$ for $s \in S - S'$ and $s' \mapsto [S']$ for $s' \in S'$.

If S is a standard subset of Σ, and S' a connected subset of S, then S/S' is a standard set obtained from S by gluing together the elements of S' to make a single root $[S']$. We obviously have the following.

3.7.3.7.5 Proposition

Let S be a standard subset of Σ, and S' a connected subset of S. Then an element $s \in S - S'$ is adjacent in S to some element of S' iff s is adjacent in S/S' to $[S']$. In particular, S is connected iff S/S' is connected.

3.7.3.7.6 Proposition

Let S be a connected standard subset of Σ. Then there is at most one root $b \in S$ which is adjacent to three distinct roots of S.

PROOF. We set up the context suggested by the following diagram:

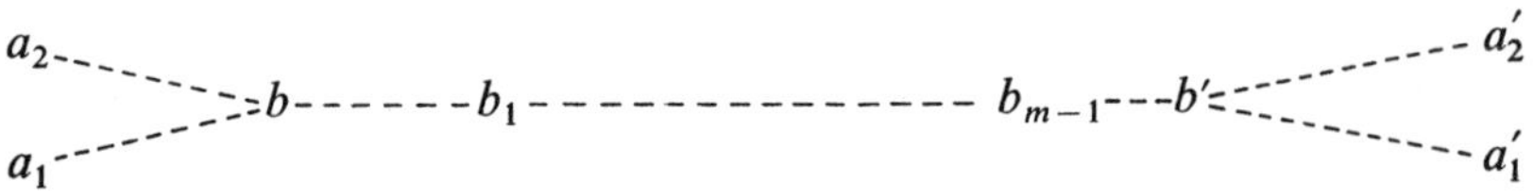

Figure 8.

Thus, suppose that b, b' are distinct roots of S and that each is adjacent to three distinct roots of S. Let $b = b_0, b_1, \ldots, b_m = b'$ be a minimal chain in S from b to b', and let $S' = \{b, b_1, \ldots, b_{m-1}, b'\}$. Since S' is connected and there are no cycles, each of b, b' is adjacent to precisely one element of S', namely b_1, b_{m-1} respectively. Let a_1, a_2 be distinct elements of $S - S'$ adjacent to b, and let a_1', a_2' be distinct elements of $S - S'$ adjacent to b'. Then a_1, a_2, a_1', a_2' are pairwise distinct, since there is no cycle. But then these are four elements of S/S' adjacent to $[S'] \in S/S'$, a contradiction in view of 3.7.3.7.3.

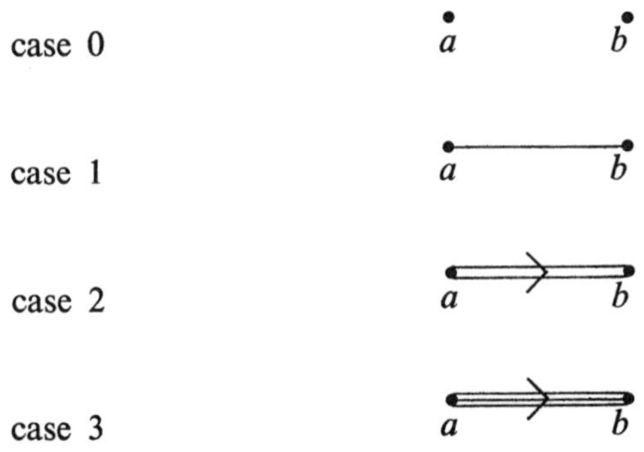

Figure 9.

Let S be a standard subset of Σ. We introduce a *diagram* D_S of S as follows. The points of D_S are points of S. Every two distinct points a, b of D_S are joined by $a^*(b)b^*(a)$ lines. (Thus by 0, 1, 2 or 3 lines, according to the table of 3.7.3.3). If one of the two roots a, b is longer than the other, this is indicated by an arrow from the longer toward the shorter root. If $S = \{a, b\}$ and $|a| \geq |b|$, then one reads off the diagram D_S of S from the table of 3.7.3.3.1 (see Figure 9).

Clearly, D_S determines $a^*(b)$, $b^*(a)$ for $a, b \in S$, and conversely.

3.7.3.7.7 Definition

If π is a simple system of Σ, then the diagram D_π of π is called the *Dynkin diagram* of Σ with respect to π.

Since any two simple systems of Σ are conjugate under W_Σ, we can speak of the *Dynkin diagram* of Σ.

3.7.3.7.8 Definition

We say that D_S is *connected* if S is connected (that is, if D_S without the arrows is connected in the usual topological sense). The *connected components* of D_S are the D_{S_i}, where the S_i are the connected components of S. If T is a second standard subset, an *isomorphism* from D_S to D_T is a bijection f from S to T which preserves the orientation (arrows) and number of lines joining pairs of points of D_S.

One sees from 3.7.3.3.1 that f is an isomorphism from D_S to D_T iff f is a bijection from S to T such that $a^*(b) = (af)^*(bf)$ for $a, b \in S$.

We shall show that for any connected standard subset of n elements, D_S is one of the diagrams in the following table. The *type* of S is defined by Figure 10.

Recall that for a standard set S and connected subset S' of S, $S/S' = (S - S') \cup \{[S']\}$. Since there are no cycles, an element $a \in S - S'$ is adjacent to $[S']$ iff a is adjacent to a unique element a' of S'.

3.7.3.7.9 Proposition

Let S' be a connected subset of a standard subset S of Σ. Suppose that

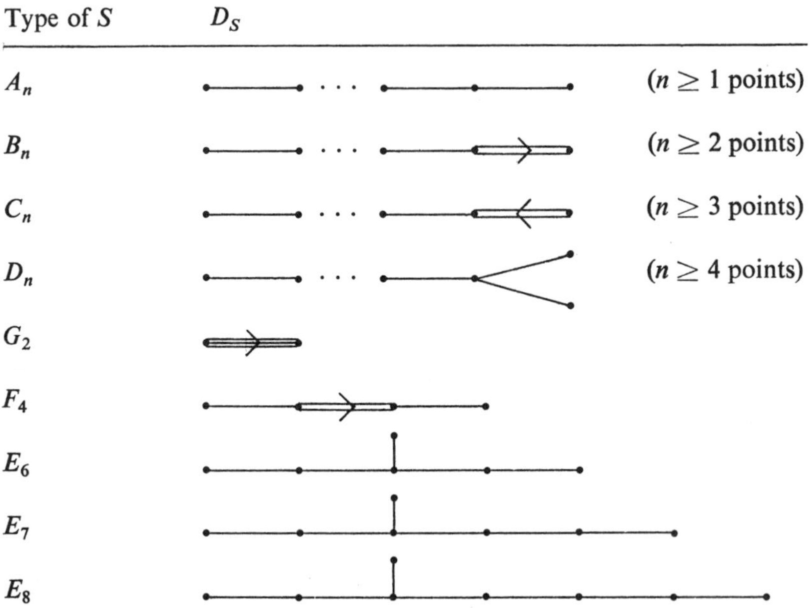

Figure 10.

$D_{S'}$ contains no double or triple line. Then $D_{S/S'}$ is obtained from D_S by shrinking S' to the point $[S']$. For example:

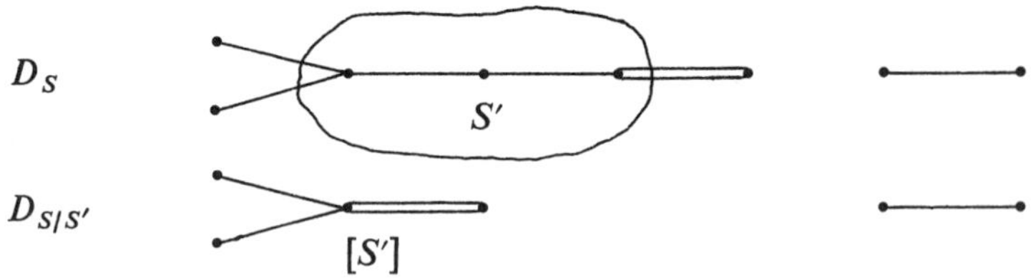

Figure 11.

PROOF. We have observed that an element a of $S - S'$ is adjacent to $[S']$ iff a is adjacent to a unique element a' of S'. The given condition on S' implies that the lengths of elements of S' are all equal. Thus, the length of $[S']$ is the same as the length of a', by 3.7.3.5.7. Now $(a, a') = (a, [S'])$, so that $a'^*(a) = [S']^*(a)$ and $a^*(a') = a^*([S'])$. Thus, the orientation and number of lines joining a and a' in D_S is the same as for a and $[S']$ in

$D_{S/S'}$. (Of course, this number is 1 or 0 and there is no orientation in this case.)

We now strengthen 3.7.3.7.6.

3.7.3.7.10 Proposition
Let S be a standard subset of Σ, and let $b \in S$. Then the total number of lines joining b to adjacent points of S is at most three.

PROOF. Let $a_1, \ldots, a_n$ be the distinct roots in S adjacent to b. Since there is no cycle in S, $(a_i, a_j) = 0$ for $i \neq j$. It follows that the r_{a_i} commute and that the root

$$c = b\prod_1^n r_{a_i}$$

is $c = b - \sum_1^n a_i^*(b)a_i$. Now

$$b^*(c) = b^*(b) - \sum_1^n a_i^*(b)b^*(a_i) = 2 - \sum n_i,$$

where $n_i = a_i^*(b)b^*(a_i)$ is the number of lines joining a_i and b in D_S. But $|b^*(c)| \leq 1$ since b and c have the same length, by 3.7.3.3.1. Thus, $\sum n_i \leq 3$.

In the next two theorems, we follow a convention for describing a vector $v = \sum_{a \in S} c_a a$, with S a standard subset of Σ, whereby the coefficients c_a are written next to the points of D_S. For example

$v =$

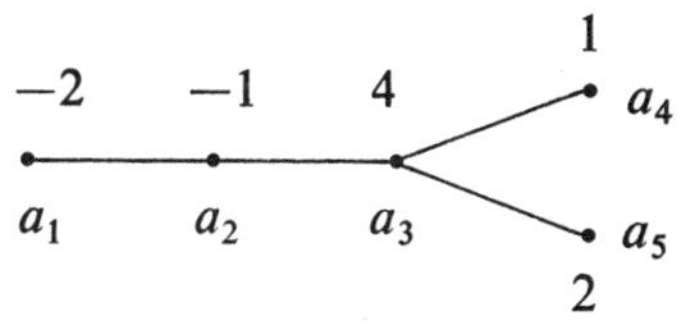

Figure 12.

is the vector $v = -2a_1 - a_2 + 4a_3 + a_4 + 2a_5$.

3.7.3.7.11 Theorem

Let S be a connected standard subset of Σ of n elements. Suppose that no double or triple lines occur in the diagram D_S. Then D_S is isomorphic to one of the following diagrams:

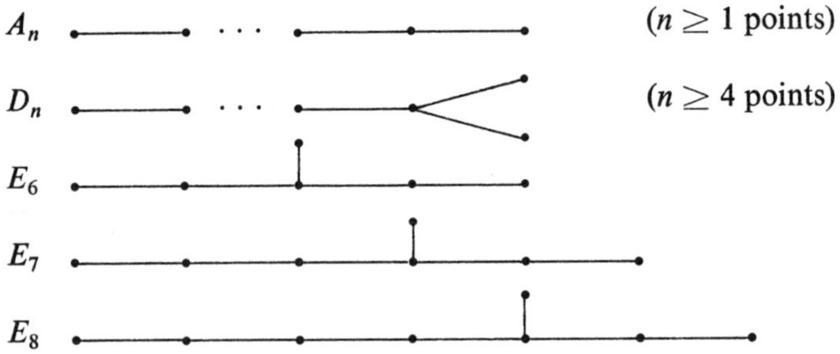

Figure 13.

PROOF. Since S is connected and adjacent points of D_S joined by a single line, the lengths of the elements of S are equal. We may assume that the lengths are 1, by replacing (,) by a scalar multiple of (,). Suppose that D_S is not isomorphic to A_n. Then there is a unique element d of S adjacent to three elements of S, by 3.7.3.7.6, so that D_S has the following form:

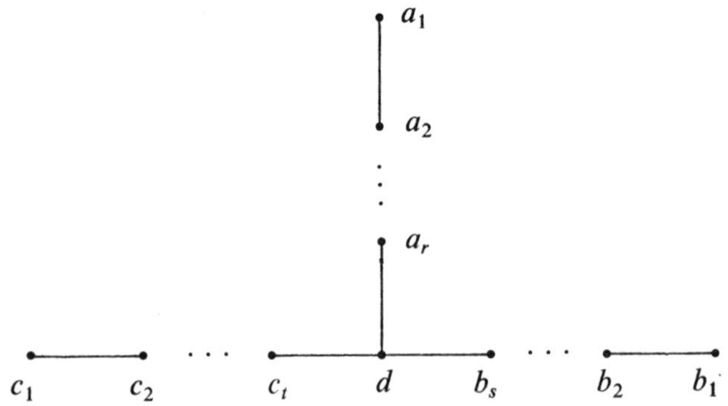

Figure 14.

Here, we may take $1 \leq r \leq s \leq t$. We now exclude three minimal diagrams, by finding in each case a nonexistent nonzero vector v with $(v, v) = 0$.

The diagram

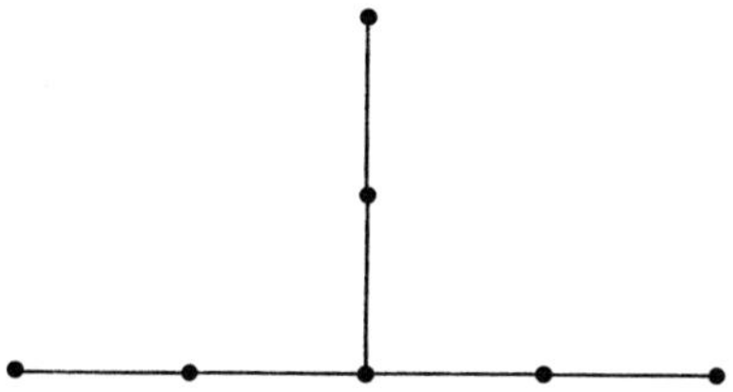

Figure 15.

cannot occur, for the vector

$v =$

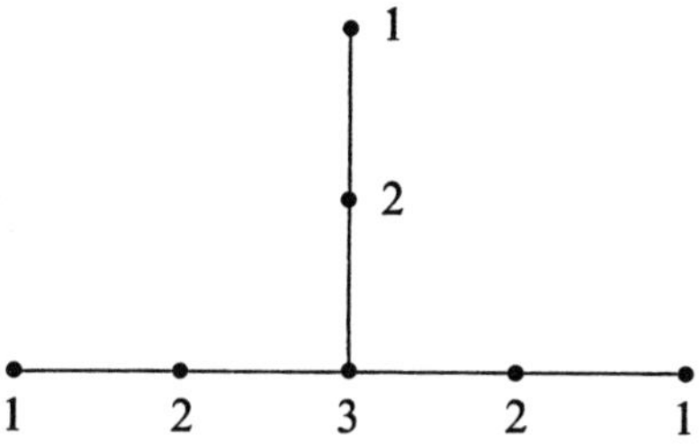

Figure 16.

would have length 0:

$$
\begin{aligned}
(v, v) &= (1 + 2^2 + 3^2) + (1 + 2^2) + (1 + 2^2) + (-2 - 6) + (-2 - 6) \\
&\quad + (-2 - 6) \\
&= 0.
\end{aligned}
$$

The diagram

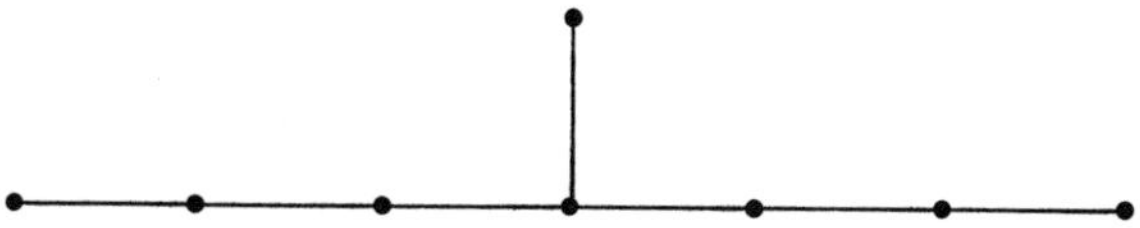

Figure 17.

does not occur, for the vector

$v =$

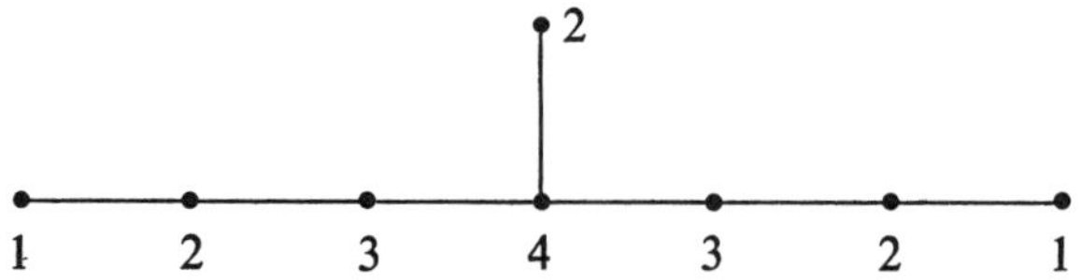

Figure 18.

would have length 0:

$$(v, v) = (1 + 2^2 + 3^2 + 4^2) + (2^2) + (3^2 + 2^2 + 1) + (-2 - 6 - 12) + (-8) + (-12 - 6 - 2) = 0.$$

The diagram

Figure 19.

does not occur, since

$v =$

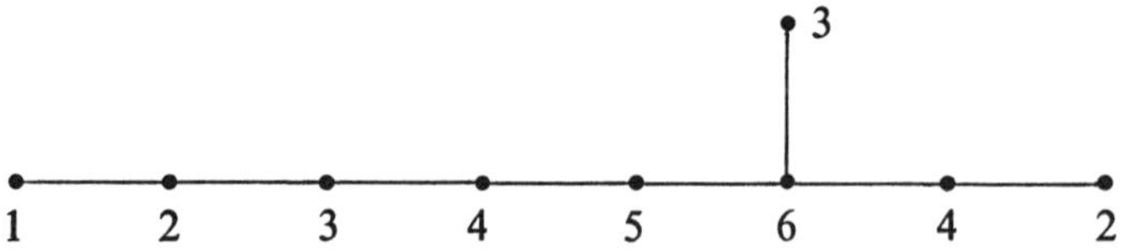

Figure 20.

would have length 0:

$$(v, v) = (1 + \ldots + 6^2) + (3^2) + (4^2 + 2^2) + (-2 - 6 - 12 - 20 - 30) + (-18) + (-24 - 8) = 0.$$

We now describe restrictions on the integers r, s, t in the preceding general diagram. Recall that $1 \leq r \leq s \leq t$. If r were greater than 1, then

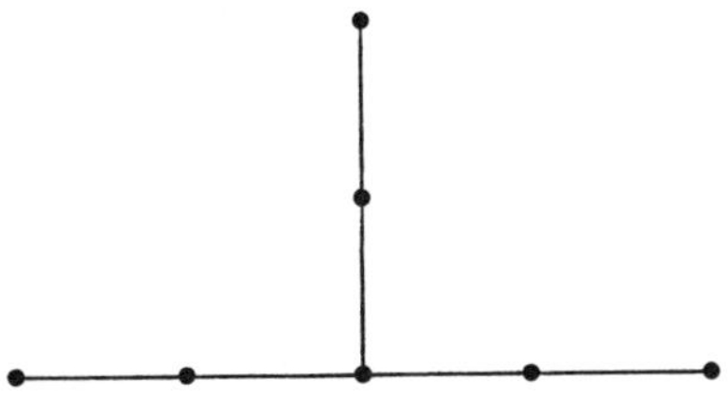

Figure 21.

would occur as the diagram of a connected subset of S, but this was ruled out above. Thus, $r = 1$. If s were greater than 2, then

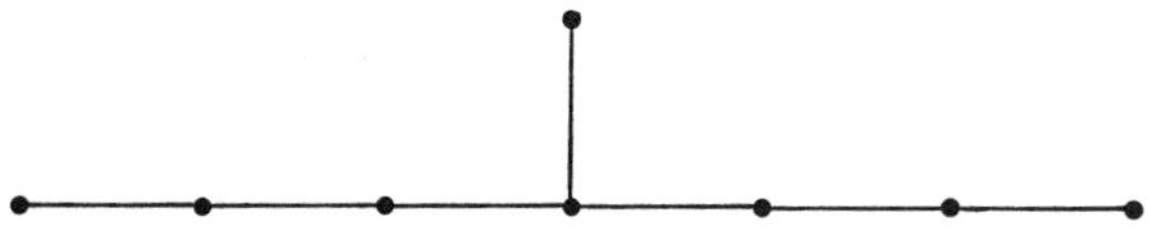

Figure 22.

would occur as diagram of a subset of S, but this was also ruled out. Thus, $s \leq 2$. For $s = 1$, D_S is of type D_n. Otherwise, $s = 2$. If t were greater than 4, then

Figure 23.

would occur as diagram of a subset of S, but this was ruled out. Thus, $t \leq 4$ for $s = 2$ and the diagram in this case is of type E_6, E_7, or E_8.

3.7.3.7.12 Theorem

Let S be a connected standard subset of n elements. Suppose that a double or triple line occurs in D_S. Then D_S is isomorphic to one of the following diagrams:

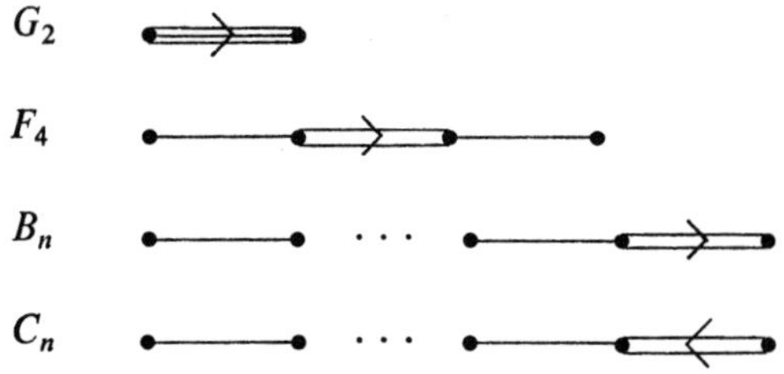

Figure 24.

PROOF. We suppress the arrows whenever they are irrelevant. If D_S contains a triple line, then D_S is of type G_2, by 3.7.3.7.10. Now suppose that D_S contains no triple line. By 3.7.3.7.5 and 3.7.3.7.9, D_S can contain no subdiagram of the form

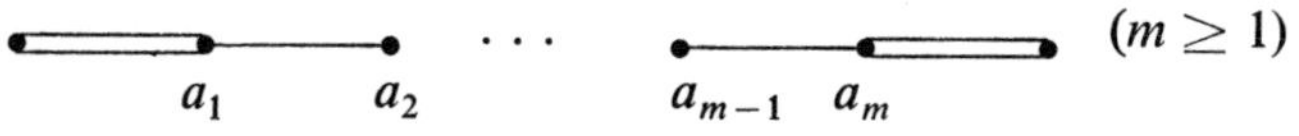

Figure 25.

or

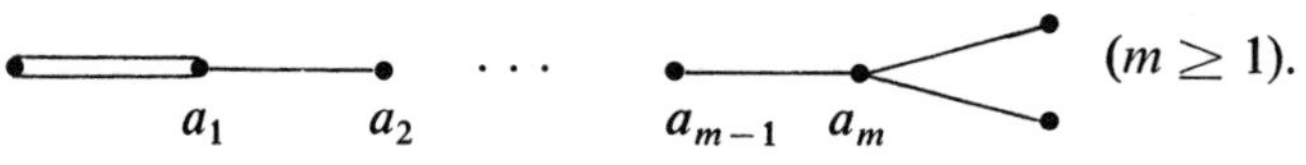

Figure 26.

For otherwise, letting $S' = \{a_1, \ldots, a_m\}$, $D_{S/S'}$ would contain a subdiagram of the form

Figure 27.

or

Figure 28.

a contradiction in view of 3.7.3.7.10. It follows either that D_S is of type F_4, B_n, or C_n, or that D_S contains a subdiagram of one of the following two kinds:

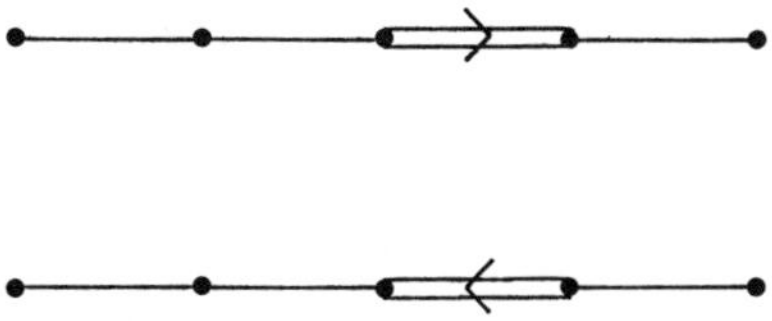

Figure 29.

But these diagrams cannot occur since the following vectors would have length 0:

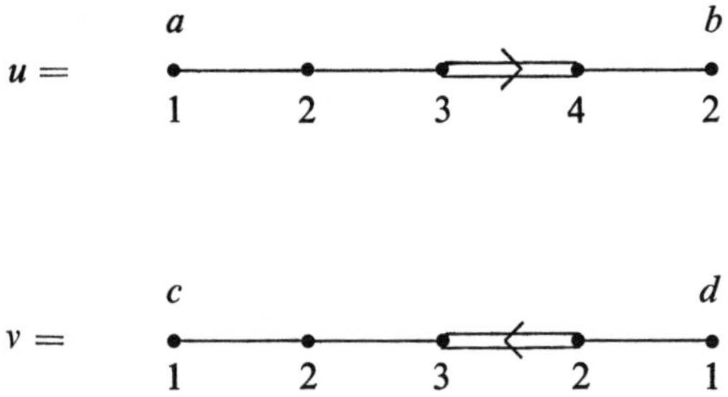

Figure 30.

for

$$\frac{(u, u)}{(b, b)} = 2(1 + 2^2 + 3^2 - 2 - 6 - 12) + 1(4^2 + 2^2 - 8) = 0,$$

$$\frac{(v, v)}{(d, d)} = 1(1 + 2^2 + 3^2 - 2 - 6) + 2(2^2 + 1^2 - 6 - 2) = 0.$$

The preceding two theorems give the classification of connected standard subsets of Σ. Each type actually occurs. This is discussed briefly in 3.7.5.

3.7.3.8 Classification of Root Systems

Recall that the Dynkin diagram of Σ with respect to a simple system π

of Σ is D_π, and that the Weyl group W_Σ acts transitively on the D_π so that the D_π are all isomorphic. The isomorphism class of the D_π is the *Dynkin diagram* of Σ. The D_π have been classified in 3.7.3.7 in the case that π is connected, so that we already know that the Dynkin diagram of an irreducible root system is of one of the types described in the table at the beginning of 3.7.3.7. Furthermore, each of these types occurs as the type of some Dynkin diagram (see 3.7.5). In general, let $\pi = \bigcup \pi_i$ where the π_i are the connected components of π. Then the Dynkin diagram D_π of Σ with respect to π is $D_\pi = \bigcup D_{\pi_i}$. The points of D_π are the points of the D_{π_i}. Two points are not connected by lines if they belong to distinct D_{π_i}, D_{π_j}. Two points belonging to D_{π_i} are oriented and connected by lines in D_{π_i} just as in D_π. These observations, together with the following theorem, give the classification of all root systems.

3.7.3.8.1 Theorem

Let (V, Σ), (V', Σ') be root systems with simple systems π, π' respectively. Then an isomorphism f from D_π to $D_{\pi'}$ can be extended to an isomorphism from (V, Σ) to (V', Σ') which maps π to π'. Conversely, the restriction $f|_\pi$ of an isomorphism f from (V, Σ) to (V', Σ') mapping π to π' is an isomorphism from D_π to $D_{\pi'}$.

PROOF. This is a restatement of 3.7.3.4.3, since, as we noted in 3.7.3.7.8, a bijection f from π to π' is an isomorphism from D_π to $D_{\pi'}$ iff $a^*(b) = (af)^*(bf)$ for $a, b \in \pi$.

3.7.3.9 Weak Root Systems

We now assume only that (V, Σ) is a weak root system, and carry over the language introduced in the discussion of root systems. Our objective is the classification of weak root systems. These turn out to be the reduced and nonreduced root systems of [7] for rank ≥ 2 and the weak root systems described below in 3.7.3.9.1 for rank $= 1$.

The classification of weak root systems is carried out by showing that $(V, \Sigma_{\min})$ is a root system and by comparing (V, Σ) and $(V, \Sigma_{\min})$. Here $\Sigma_{\min} = \{a \in \Sigma \mid a \text{ is minimal}\}$, where a is said to be *minimal* if $|c| \geq 1$ whenever ca is root.

3.7.3.9.1 Example

Let $V = \mathbb{Q}$ together with the form $(v, w) = vw$, and let $\Sigma = \{\pm 1, \pm 2, \ldots, \pm n\}$. Then (V, Σ) is a weak root system. The minimal roots are ± 1.

3.7.3.9.2 Proposition

For each $a \in \Sigma$, there is a minimal root a_0 and an integer n such that

$$\mathbb{Q}a_0 \cap \Sigma = \{\pm ia_0 \mid 1 \leq i \leq n\} \text{ and } a = ia_0 \text{ for some } i.$$

PROOF. This is clear from the arguments used in proving 3.7.3.1.6.

3.7.3.9.3 Proposition

Let π be a simple system of Σ. Then π is a basis for V and each element b of Σ may be written $b = \pm \sum_{a \in \pi} c_a a$ where the c_a are nonnegative integers for $a \in \pi$.

PROOF. The proof of 3.7.3.2.2 applies here as well.

3.7.3.9.4 Proposition

Let $a \in \Sigma_{\text{min}}$. Then $a \in \pi$ for some simple system π.

PROOF. The proof of 3.7.3.2.3 applies here under the present assumption that a is minimal.

3.7.3.9.5 Proposition

$a^*(b) \in \mathbb{Z}$ for $a \in \Sigma_{\text{min}}$, $b \in \Sigma$.

PROOF. The proof of 3.7.3.3.1 applies here.

3.7.3.9.6 Theorem

(V, Σ_{min}) is a root system.

PROOF. Since the r_a are automorphisms of (V, Σ), Σ_{min} is stable under the r_a. And V is spanned by Σ_{min}, by 3.7.3.9.2. It remains to show that for $a, b \in \Sigma_{\text{min}}$ and $(a, b) < 0$, $a + b \in \Sigma_{\text{min}} \cup \{0\}$. Thus, let $a, b \in \Sigma_{\text{min}}$ with $(a, b) < 0$ and $a + b \neq 0$. Since a and b are minimal, they are linearly independent, by 3.7.3.9.2. We may take $|a| \geq |b|$. The table of 3.7.3.3 now applies to a, b, since $a^*(b)$ and $b^*(a)$ are integers. Thus, $a^*(b) = -1$. But then $br_a = b - a^*(b)a = b + a$ and $b + a \in \Sigma_{\min}$.

A weak root system is completely determined by giving $\Sigma_{\min}$ and giving $f(a)$ for $a \in \Sigma_{\min}$, where $f(a)$ is the greatest integer n such that na is a root (see 3.7.3.9.2). Thus, by 3.7.3.9.4, it is determined by the Dynkin diagram D of $\Sigma_{\min}$ where each vertex a of D is labeled by $f(a)$. This labeled diagram is called the *diagram* of Σ. We leave it to the reader to verify the following, by using 3.7.3.1.1 and 3.7.3.9.2 and looking at A_2, B_2, G_2:

1. if two vertices are joined by a single line, both are labeled by 1;
2. if two vertices a, b are joined by a double line with a long, then a is labeled by 1 and b by 1 or 2;
3. if two vertices are joined by a triple line, then both are labeled by 1.

The reader can now easily prove the following theorem, which classifies the weak root systems.

3.7.3.9.7 Theorem

Let Σ be a weak root system which is not a root system. Suppose that the diagram of Σ is connected. Then either Σ is the weak root system of rank 1 described in 3.7.3.9.1 or $\Sigma = \Sigma_{\min} \cup \{2a \mid a \text{ is a short root of } \Sigma_{\min}\}$ and $\Sigma_{\min}$ is a root system of type B_n. In the latter case, the diagram of Σ is of type B_n and the short root is labeled by 2.

3.7.4 Isomorphisms of Irreducible Modules and Semisimple Lie Algebras

We prove in this section that two split irreducible modules for a split semisimple Lie algebra $\mathfrak{L}$ of characteristic 0 are isomorphic iff they have "the same highest weight." We then use a similar argument to show that two split semisimple Lie algebras of characteristic 0 are isomorphic iff "their root systems are isomorphic." The existence of split irreducible modules and split semisimple Lie algebras of characteristic 0 is discussed briefly in the next section.

We now assume that $\mathfrak{L}$ is a split semisimple Lie algebra of characteristic

0, $\mathfrak{T}$ a split Cartan subalgebra of $\mathfrak{L}$, Σ the root system of $\mathfrak{L}$ with respect to $\mathfrak{T}$, π a simple system of Σ, Σ^+ the corresponding half-system, $\Sigma^- = \Sigma - \Sigma^+, \mathfrak{L}^+ = \sum_{a \in \Sigma^+} \mathfrak{L}_a$, and $\mathfrak{L}^- = \sum_{a \in \Sigma^-} \mathfrak{L}_a$. Here, $\mathfrak{L}_a = \mathfrak{L}_a(\mathfrak{T})$ for $a \in \mathfrak{T}^*$. For $a \in \Sigma$, we let t_a be the element of $\mathfrak{T}$ such that $b(t_a) = (a, b)$ for all $b \in \mathfrak{T}^*$, choose $e_a \in \mathfrak{L}_a, f_a \in \mathfrak{L}_{-a}$ such that $e_a f_a = t_a$, and let $\mathfrak{L}^{(a)}$ be the split simple Lie algebra with basis e_a, t_a, f_a, namely, $\mathfrak{L}^{(a)} = \mathfrak{L}_a + \mathfrak{T}_a + \mathfrak{L}_{-a}$.

Next, we let $\mathfrak{M}$ be a nonzero $\mathfrak{L}$-module *split by* $\mathfrak{T}$. By split, we mean that $\mathfrak{M} = \sum_{b \in \mathfrak{T}^*} \oplus \mathfrak{M}_b(\mathfrak{T})$. Let f be the representation of $\mathfrak{L}$ afforded by $\mathfrak{M}$.

3.7.4.1 Definition

An element x of $\mathfrak{M}$ is *extreme* if $x \neq 0$ and $x\mathfrak{L}^+ = \{0\}$.

3.7.4.2 Proposition

For some $b \in \mathfrak{T}^*$, $\mathfrak{M}_b(\mathfrak{T})$ contains an extreme element x.

PROOF. We show first that for $a \in \mathfrak{L}^+, f(a)$ is nilpotent. Let $C = \{\sum_{a \in \pi} c_a a \mid c_a \in \mathbf{Z}, c_a \geq 0 \text{ for } a \in \pi\}$ and let $L(b) = \sum c_a$ be the level of $b = \sum c_a a \in C$. For $d \in \mathfrak{T}^*$, let $\mathfrak{M}_i^d = \sum_{\substack{b \in C \\ L(b) \geq i}} \mathfrak{M}_{d+b}(\mathfrak{T})$. Then $\mathfrak{M}_d(\mathfrak{T}) f(a) \subset \mathfrak{M}_1^d$ and $\mathfrak{M}_i^d f(a) \subset \mathfrak{M}_{i+1}^d$ for all $i \geq 1$, $d \in \mathfrak{T}^*$. Letting $n = \max L(b)$ where b ranges over all elements b of $\mathfrak{T}^*$ such that $d + b$ is a weight of $\mathfrak{T}$ in $\mathfrak{M}$ for some weight d of $\mathfrak{T}$ in $\mathfrak{B}$, then $\mathfrak{M}_{n+1}^d = \{0\}$ for all d. Thus $f(a)^{n+1} = 0$ for $a \in \mathfrak{L}^+$. It follows from Engel's theorem that $\mathfrak{M}$ contains $v \neq 0$ such that $v\mathfrak{L}^+ = \{0\}$. Letting $v = \sum_{d \in \mathfrak{T}^*} v_d$ with $v_d \in \mathfrak{M}_d(\mathfrak{T})$ for all d, we can choose a particular d such that $v_d \neq 0$. Let $x = v_d$ for such a d. For $a \in \Sigma^+$, $0 = ve_a = \Sigma v_d e_a$. Since $v_d e_a$ is the $(d + a)$th component of 0, $xe_a = 0$ for $a \in \Sigma^+$ and $x\mathfrak{L}^+ = \{0\}$.

We wish to study the submodule of $\mathfrak{M}$ generated by an extreme element contained in some $\mathfrak{M}_b(\mathfrak{T})$. For this, we need the following.

3.7.4.3 Proposition

$\mathfrak{L}$ is generated as a Lie algebra by $S = \{e_a, f_a, t_a \mid a \in \pi\}$.

PROOF. Let $\mathfrak{B}$ be the Lie subalgebra generated by S. For $a \in \Sigma$, let $w_a \in \operatorname{Aut} \mathfrak{L}$ be defined as in 3.7.2.4, and note that w_a stabilizes $\mathfrak{B}$ for $a \in \pi$. Let W_π be the subgroup of Aut $\mathfrak{L}$ generated by $\{w_a \mid a \in \pi\}$ and W_π^* the subgroup of Aut Σ generated by $\{w_a^* \mid a \in \pi\}$, and recall the homomorphism $w \mapsto w^*$ from W_π onto W_π^*. By 3.7.2.7, W_π^* is the Weyl group of Σ. Thus, for any $b \in \Sigma$, there exists $a \in \pi$ and $w^* \in W_\pi^*$ such that $aw^* = b$, by 3.7.3.2.3 and 3.7.3.4.2. But then $(\mathfrak{L}_a)^w = \mathfrak{L}_{aw*} = \mathfrak{L}_b$ and $\mathfrak{L}_b \subset \mathfrak{B}^w = \mathfrak{B}$. Thus, $\mathfrak{L} = \mathfrak{T} + \sum_{b \in \Sigma} \mathfrak{L}_b \subset \mathfrak{B}$ and $\mathfrak{L} = \mathfrak{B}$.

3.7.4.4 Definition

For $v \in \mathfrak{M}$ and $a_1, \ldots, a_n \in \mathfrak{L}$, $[v, a_1, \ldots, a_n] = (\ldots(va_1)\ldots a_{n-1})a_n$. If $n = 0$, $[v] = v$.

3.7.4.5 Proposition

Let x be an extreme element of $\mathfrak{M}_b(\mathfrak{T})$. Then the $\mathfrak{L}$-submodule $\mathfrak{W}$ generated by x is spanned by the elements of the form $[x, f_{a_1}, \ldots, f_{a_n}]$ where the a_i are from π and $n \geq 0$. Moreover, $\mathfrak{W}_b(\mathfrak{T}) = kx$ and b is the unique weight of $\mathfrak{T}$ on $\mathfrak{W}$ such that every other weight of $\mathfrak{T}$ on $\mathfrak{W}$ is of the form $b - \sum_{a \in \pi} c_a a$ with $c_a \in \mathbf{Z}$, $c_a \geq 0$ for $a \in \pi$.

PROOF. Let $\mathfrak{W}_0$ be the span of the elements of the form $[x, f_{a_1}, \ldots, f_{a_n}]$ where the $a_i \in \pi$ and $n \geq 0$. For $t \in \mathfrak{T}$,

$$[x, f_{a_1}, \ldots, f_{a_n}]t = (b(t) - \textstyle\sum a_i(t))[x, f_{a_1}, \ldots, f_{a_n}].$$

Thus, $\mathfrak{W}_0$ is $\mathfrak{T}$-stable and $\mathfrak{W}_0 \subset kx + \sum_{d \in C,\, d \neq 0} \mathfrak{M}_{b-d}(\mathfrak{T})$ where

$$C = \{\sum_{a \in \pi} c_a a \mid c_a \in \mathbf{Z},\ c_a \geq 0 \text{ for } a \in \pi\}.$$

This shows that $(\mathfrak{W}_0)_b(\mathfrak{T}) = kx$ and every weight of $\mathfrak{W}_0$ has the form $b - d$ with $d \in C$. If b' is another weight of $\mathfrak{T}$ with this property, then $b' = b - d$ and $b = b' - d'$ with $d, d' \in C$. But then $d' = -d$, which is possible only for $d = 0$, and $b' = b$. It remains to show that $\mathfrak{W} = \mathfrak{W}_0$, that is, that $\mathfrak{W}_0$ is $\mathfrak{L}$-stable. Clearly, $\mathfrak{W}_0$ is stable under the f_a $(a \in \pi)$, and under $\mathfrak{T}$ by a preceding observation. Since $\mathfrak{L}$ is generated by the e_a, t_a, f_a

$(a \in \pi)$, it remains only to show that $\mathfrak{W}_0$ is stable under the e_a $(a \in \pi)$. For $v \in \mathfrak{M}_{b'}(\mathfrak{T})$, $(vf_d)e_a = (ve_a)f_d$ for distinct a, $d \in \pi$ since $e_a f_d = 0$. And $(vf_a)e_a = (ve_a)f_a - b'(t_a)v$ for $a \in \pi$, since $e_a f_a = t_a$. Thus,

$$[x, f_{a_1}, \ldots, f_{a_n}, e_a] + \mathfrak{W}_0 = [x, f_{a_1}, \ldots, f_{a_{n-1}}, e_a, f_{a_n}] + \mathfrak{W}_0 = \ldots = [x, e_a, f_{a_1}, \ldots, f_{a_n}] + \mathfrak{W}_0 = \mathfrak{W}_0 \text{ for } a \in \pi.$$

Thus, $\mathfrak{W}_0$ is $\mathfrak{L}$-stable and $\mathfrak{W} = \mathfrak{W}_0$.

3.7.4.6 Definition

d is the *highest weight* of $\mathfrak{T}$ in $\mathfrak{M}$ with respect to π if every other weight in $\mathfrak{M}$ has the form $d - \sum_{a \in \pi} c_a a$ where $c_a \in \mathbb{Z}$, $c_a \geq 0$ for $a \in \pi$.

3.7.4.7 Theorem

Every irreducible $\mathfrak{L}$-module $\mathfrak{M}$ split over $\mathfrak{T}$ has a unique highest weight d with respect to π, and the corresponding weight space has the form $\mathfrak{M}_d(\mathfrak{T}) = kx$ where x is extreme. If $\mathfrak{M}'$ is a second irreducible $\mathfrak{L}$-module split over $\mathfrak{T}$ with highest weight d' with respect to π, then $\mathfrak{M}$ and $\mathfrak{M}'$ are $\mathfrak{L}$-isomorphic iff $d = d'$.

PROOF. Choose d such that $\mathfrak{M}_d(\mathfrak{T})$ contains an extreme element x. Then $\mathfrak{M}$ is the $\mathfrak{L}$-submodule generated by x since $\mathfrak{M}$ is $\mathfrak{L}$-irreducible. Thus, $\mathfrak{M}_d(\mathfrak{T}) = kx$ and d is the unique highest weight, by 3.7.4.5. Suppose now that $\mathfrak{M}$, $\mathfrak{M}'$ are $\mathfrak{L}$-irreducible and split over $\mathfrak{T}$ with highest weights d, d' respectively. If $\mathfrak{M}$ and $\mathfrak{M}'$ are $\mathfrak{L}$-isomorphic, then obviously $d = d'$. Suppose, conversely, that $d = d'$. Let $\mathfrak{M}_d(\mathfrak{T}) = kx$, $\mathfrak{M}'_d(\mathfrak{T}) = kx'$. Let $\mathfrak{W}$ be the $\mathfrak{L}$-submodule of $\mathfrak{M} \oplus \mathfrak{M}'$ generated by $X = x \oplus x'$. We claim that $\mathfrak{M}$ and $\mathfrak{M}'$ are both isomorphic to $\mathfrak{W}$, hence to each other. Let π, π' be the projections of $\mathfrak{M} \oplus \mathfrak{M}'$ on $\mathfrak{M}$, $\mathfrak{M}'$ respectively. Then π and π' are $\mathfrak{L}$-homomorphisms such that $\pi(\mathfrak{W}) = \mathfrak{M}$ and $\pi'(\mathfrak{W}) = \mathfrak{M}'$. Thus, it suffices to show that $\pi|_{\mathfrak{W}}$ and $\pi'|_{\mathfrak{W}}$ are injective, that is, that $\mathfrak{W} \cap (\{0\} \oplus \mathfrak{M}') = \mathfrak{W} \cap (\mathfrak{M} \oplus \{0\}) = \{0 \oplus 0\}$. In view of the irreducibility of $\mathfrak{M}$ and $\mathfrak{M}'$, the only alternative is that $\mathfrak{W}$ contain $\{0\} \oplus \mathfrak{M}'$ or $\mathfrak{M} \oplus \{0\}$. But then it would follow that $\mathfrak{W}$ contains x or x', hence x and x' since $X = x \oplus x' \in \mathfrak{W}$. But then $\mathfrak{W}$ contains $\{0\} \oplus \mathfrak{M}'$ and $\mathfrak{M} \oplus \{0\}$, and

$\mathfrak{W} = \mathfrak{M} \oplus \mathfrak{M}'$. This is impossible, for $\mathfrak{W}_d(\mathfrak{T}) = kX$ and $(\mathfrak{M} \oplus \mathfrak{M}')_d(\mathfrak{T}) \supseteq kx + kx'$. Thus, π and π' are injective and $\mathfrak{M}$, $\mathfrak{M}'$ $\mathfrak{L}$-isomorphic.

We next show that two split semisimple Lie algebras over k with isomorphic root systems are isomorphic. We begin by relating the decomposition of $\mathfrak{L}$ as the sum of its simple ideals and the decomposition of the root system (V, Σ) of $\mathfrak{L}$ as the sum of its irreducible direct factors.

3.7.4.8 Theorem

Let $(V, \Sigma) = \sum \oplus (V_i, \Sigma_i)$ be the decomposition of (V, Σ) as direct sum of its irreducible direct factors (V_i, Σ_i). Let

$$\mathfrak{L}_i = \sum_{a \in \Sigma_i} \mathfrak{L}^{(a)}, \; \mathfrak{T}_i = \sum_{a \in \Sigma_i} \mathfrak{L}_a \mathfrak{L}_{-a}.$$

Then $\mathfrak{L} = \sum \oplus \mathfrak{L}_i$ is the decomposition of $\mathfrak{L}$ as direct sum of its simple ideals $\mathfrak{L}_i$. The root system of $\mathfrak{L}_i$ with respect to $\mathfrak{T}_i$ is isomorphic to (V_i, Σ_i).

PROOF. We know that $V = \sum \oplus V_i$ (orthogonal direct sum) and $\Sigma_i = \Sigma \cap V_i$ for all i. Thus, a, $b \in \Sigma_i$ and $a + b \in \Sigma \Rightarrow a + b \in \Sigma_i$ for all i. Thus,

$$\mathfrak{L}_i = \sum_{a \in \Sigma_i} \mathfrak{L}_a \oplus \mathfrak{T}_i$$

is a subalgebra of $\mathfrak{L}$ for all i. Obviously, $\mathfrak{T}_i$ is a split Cartan subalgebra of $\mathfrak{L}_i$. We also know that $a + b \notin \Sigma_i$ for $a \in \Sigma_i$, $b \in \Sigma_j$ and $i \neq j$, by 3.7.3.6.6. Thus, $\mathfrak{L}_a \mathfrak{L}_b = \{0\}$ for such a, b. Since $\mathfrak{L}_b t_a = \{0\} = \mathfrak{L}_a t_b$ for $(a, b) = 0$, we therefore have $\mathfrak{L}_i \mathfrak{L}_j = \{0\}$ for $i \neq j$. Thus, the $\mathfrak{L}_i$ are ideals, hence semisimple as Lie algebras, by 2.7.4 and 2.8.5. The root system of $\mathfrak{L}_i$ with respect to $\mathfrak{T}_i$ is (V'_i, Σ'_i) where $V'_i = \mathfrak{T}^*_{i\mathrm{Q}}$ and $\Sigma'_i = \{a|_{T_i} \mid a \in \Sigma_i\}$. Consider the linear mapping $a \mapsto a|_{\mathfrak{T}_i}$ from $V = \mathfrak{T}^*_{\mathrm{Q}}$ onto $V'_i = \mathfrak{T}^*_{i\mathrm{Q}}$. Its kernel is $W = \{a \in V \mid a|_{\mathfrak{T}_i} = 0\} = \{a \in V \mid 0 = a(t_b) = (a, b) \text{ for } b \in \Sigma_i\} = \{a \in V \mid a \perp V_i\} = V_i^{\perp}$. Since $V_i \perp V_j$ for $i \neq j$, $W = V_i^{\perp} = \sum_{j \neq i} \oplus V_j$, by the nondegeneracy of (,). Thus, $a \mapsto a|_{\mathfrak{T}_i}$ induces by restriction a linear isomorphism from V_i onto V'_i mapping Σ_i onto Σ'_i, and (V_i, Σ_i) and (V'_i, Σ'_i) are isomorphic.

It remains to show that the $\mathfrak{L}_i$ are simple. Since the root system of $\mathfrak{L}_i$ is irreducible, this amounts to showing that $\mathfrak{L}$ is simple if (V, Σ) is irreducible, or that (V, Σ) is not irreducible if L is not simple. Thus, let $\mathfrak{L} = \mathfrak{L}^1 \oplus \mathfrak{L}^2$ where $\mathfrak{L}^i$ is a proper ideal of $\mathfrak{L}$ $(i = 1, 2)$. Then $\mathfrak{T} = \mathfrak{T}^1 \oplus \mathfrak{T}^2$ and $\Sigma = \Sigma^1 \cup \Sigma^2$ where $\mathfrak{L}^i = \mathfrak{T}^i \oplus \sum_{a \in \Sigma^i} \mathfrak{L}_a$ $(i = 1, 2)$. Now $\mathfrak{L}^1 \perp \mathfrak{L}^2$ relative to K(,), by 2.8.5, so that $\mathfrak{T}^1 \perp \mathfrak{T}^2$. But $\mathfrak{T}^i = \sum_{a \in \Sigma^i} \mathfrak{L}_a \mathfrak{L}_{-a}$ is the span of $\{t_a \mid a \in \Sigma^i\}$, so that $\Sigma^1 \perp \Sigma^2$. Thus, $V = V^1 \oplus V^2$ where V^i is the span of Σ^i in V $(i = 1, 2)$, and (V, Σ) is not irreducible since (V^i, Σ^i) is a proper direct factor $(i = 1, 2)$.

We shall later prove that for two split Cartan subalgebras $\mathfrak{H}_1$, $\mathfrak{H}_2$ of $\mathfrak{L}$, there is an automorphism f of $\mathfrak{L}$ such that $f(\mathfrak{H}_1) = \mathfrak{H}_2$. It follows that any two root systems for $\mathfrak{L}$ are isomorphic. This observation clarifies (but is not technically necessary for) the following theorem.

3.7.4.9 Theorem

Two split semisimple Lie algebras $\mathfrak{L}$, $\mathfrak{L}'$ over k are isomorphic iff they have isomorphic root systems. If $\mathfrak{T}$, $\mathfrak{T}'$ are split Cartan subalgebras of $\mathfrak{L}$, $\mathfrak{L}'$ respectively and if $a \mapsto a'$ is an isomorphism from the root system (V, Σ) of $\mathfrak{L}$ with respect to $\mathfrak{T}$ to the root system (V', Σ') of $\mathfrak{L}'$ with respect to $\mathfrak{T}'$, then any linear mapping f from $\sum_{a \in \pi} \mathfrak{L}_a$, π a given simple system of Σ, into $\mathfrak{L}'$ such that $(\mathfrak{L}_a)f = \mathfrak{L}_{a'}$ for $a \in \pi$ can be extended to an isomorphism f from $\mathfrak{L}$ to $\mathfrak{L}'$ such that $(\mathfrak{L}_a)f = \mathfrak{L}_{a'}$ for $a \in \Sigma$. Such an f maps $\mathfrak{T}$ to $\mathfrak{T}'$.

PROOF. We may assume without loss of generality that $\mathfrak{L}$ and $\mathfrak{L}'$ are simple, by 3.7.4.8.

If $\mathfrak{L}$ and $\mathfrak{L}'$ are isomorphic, they obviously have isomorphic root systems.

Conversely, we show that $\mathfrak{L}$ and $\mathfrak{L}'$ are isomorphic if $\mathfrak{T}$, $\mathfrak{T}'$ are split Cartan subalgebras of $\mathfrak{L}$, $\mathfrak{L}'$ respectively, (V, Σ), (V', Σ') are the root systems of $\mathfrak{L}$, $\mathfrak{L}'$ with respect to $\mathfrak{T}$, $\mathfrak{T}'$ respectively, and there is an isomorphism $a \mapsto a'$ from (V, Σ) to (V', Σ'). In this, we may assume that $(a, b) = (a', b')'$ for $a, b \in \Sigma$. Let π be a simple system for Σ, $\pi' = \{a' \mid a \in \pi\}$. Let Σ^+, Σ'^+ be the half-systems of Σ, Σ' corresponding to π, π', respectively. Regarding $\mathfrak{L}$, $\mathfrak{L}'$ as Lie modules for $\mathfrak{L}$, $\mathfrak{L}'$ respectively, through the adjoint representations, $\mathfrak{L}$ and $\mathfrak{L}'$ are irreducible modules by the simplicity of $\mathfrak{L}$, $\mathfrak{L}'$. We now mimic the proof of 3.7.4.7. Let $d \in \Sigma$ be the highest weight of $\mathfrak{L}$ with respect to π. Then obviously $d' \in \Sigma'$ is the highest weight of $\mathfrak{L}'$ with respect to π', since in the present setting the weights are the roots and the root systems isomorphic. Let $\mathfrak{L}_d = \mathfrak{L}_d(\mathfrak{T}) = kx$, $\mathfrak{L}'_d = \mathfrak{L}'_{d'}(\mathfrak{T}') = kx'$ and let $X = x \oplus x'$ in the semisimple Lie algebra $\mathfrak{L} \oplus \mathfrak{L}'$. We now construct a "diagonal subalgebra" $\mathfrak{B}$ of $\mathfrak{L} \oplus \mathfrak{L}'$ and show it to be isomorphic via the projections to both $\mathfrak{L}$ and $\mathfrak{L}'$. The key step is to show that $\mathfrak{B}$ is proper in $\mathfrak{L} \oplus \mathfrak{L}'$, which amounts to showing that the $\mathfrak{B}$-submodule $\mathfrak{W}$ of $\mathfrak{L} \oplus \mathfrak{L}'$ generated by X is proper in $\mathfrak{L} \oplus \mathfrak{L}'$. For $a \in \Sigma$, choose $e_a \in \mathfrak{L}_a, f_a \in \mathfrak{L}_{-a}, e_{a'} \in \mathfrak{L}'_{a'}, f_{a'} \in \mathfrak{L}'_{-a'}$ such that $e_a f_a = t_a$, $e_{a'} f_{a'} = t_{a'}$ for $a \in \Sigma$. Recall that $b(t_a) = (a, b) = (a', b')' = b'(t_{a'})$ for all $a, b \in \Sigma$. (Here we make use of the normalization $(a, b) = (a', b')'$ introduced above.) Let $E_a = e_a \oplus e_{a'}$, $F_a = f_a \oplus f_{a'}$, $T_a = t_a \oplus t_{a'}$, in $\mathfrak{L} \oplus \mathfrak{L}'$. Let $\mathfrak{B}$ be the Lie subalgebra of $\mathfrak{L} \oplus \mathfrak{L}'$ generated by the E_a, F_a, T_a where a ranges over π. Let $\mathfrak{W}$ be the span of the set $\{[X, F_{a_1}, F_{a_2}, \ldots, F_{a_n}] \mid a_i \in \pi, n \geq 0\}$. Using the condition $(a, b) = (a', b')'$, one easily checks that

$$[X, F_{a_1}, F_{a_2}, \ldots, F_{a_n}]T_a = ((d, a) - \Sigma(a_i, a))[X, F_{a_1}, F_{a_2}, \ldots, F_{a_n}]$$

for $a_i \in \pi (1 \leq i \leq n)$ and $n \geq 0$. Since $E_a F_b = 0$ for $a \neq b$ and $E_a F_a = T_a$, it therefore follows as in the proof of 3.7.4.7 that $\mathfrak{W}$ is $\mathfrak{B}$-stable. Now $\mathfrak{W} \neq \mathfrak{L} \oplus \mathfrak{L}'$, since

$$\mathfrak{W} \subset kx + \sum_{\substack{a \neq d \\ b' \neq d'}} (\mathfrak{L}_a \oplus \mathfrak{L}_b') \subsetneqq \mathfrak{L} \oplus \mathfrak{L}'.$$

It follows that $\mathfrak{B} \neq \mathfrak{L} \oplus \mathfrak{L}'$. For otherwise $\mathfrak{W}$ would be a proper ideal of $\mathfrak{L}$, whereas it cannot be because the only proper ideals of $\mathfrak{L}$ are $\mathfrak{L} \oplus \{0\}$ and $\{0\} \oplus \mathfrak{L}$, by 2.7.4. We now can show that $\mathfrak{L}$ and $\mathfrak{L}'$ are isomorphic to $\mathfrak{B}$, hence to each other. For the projections π, π' from $\mathfrak{L} \oplus \mathfrak{L}'$ onto

$\mathfrak{L}$, $\mathfrak{L}'$ respectively are Lie algebra homomorphisms, and $\pi|_{\mathfrak{B}}$, $\pi'|_{\mathfrak{B}}$ are injective. The injectivity is proved as in the proof of 3.7.4.7. If, say, $\pi|_{\mathfrak{B}}$ were not injective, then $\mathfrak{B} \cap \{0\} \oplus \mathfrak{L}' \neq \{0\}$. By the simplicity of $\mathfrak{L}'$, it would follow that $\mathfrak{B} \supset \{0\} \oplus \mathfrak{L}'$, hence that $x' \in \mathfrak{B}$, hence that $x = X - x' \in \mathfrak{B}$. But then $\mathfrak{B} \supset x(\mathfrak{L} \oplus \mathfrak{L}') = \mathfrak{L} \oplus \{0\}$ and $\mathfrak{B} = \mathfrak{L} \oplus \mathfrak{L}'$, a contradiction.

The isomorphisms from $\mathfrak{L}$, $\mathfrak{L}'$ to $\mathfrak{B}$ described above induce an isomorphism f from $\mathfrak{L}$ to $\mathfrak{L}'$ such that $e_a f = e_{a'}$, $f_a f = f_{a'}$ for all $a \in \pi$. Such an isomorphism f stabilizes $\mathfrak{T}$ since $\mathfrak{T}$ is spanned by $\{t_a \mid a \in \pi\}$ and $e_a f_a = t_a$, $e_{a'} f_{a'} = t_{a'}$ imply that $t_a f = t_{a'}$ for $a \in \pi$. Now, as in 3.7.2.5, $\mathfrak{L}_b f = \mathfrak{L}'_{bf*}$ for all $b \in \Sigma$. But then $af^* = a'$ for $a \in \pi$, so $bf^* = b'$ for all $b \in \Sigma$, since Σ is contained in the span of π. Thus, $\mathfrak{L}_b f = \mathfrak{L}_b'$ for all $b \in \Sigma$.

Finally, the flexibility in choosing the $e_a \in \mathfrak{L}_a$, $e_{a'} \in \mathfrak{L}_{a'}$ shows that any linear mapping f_0 from $\sum_{a\in\pi} \mathfrak{L}_a$ into $\mathfrak{L}'$ such that $\mathfrak{L}_a f_0 = \mathfrak{L}_{a'}$ for all $a \in \pi$ occurs as the restriction of f to $\sum_{a\in\pi} \mathfrak{L}_a$ if the e_a, $e_{a'}$ are suitably chosen. Thus, any such linear mapping f_0 extends to an isomorphism from $\mathfrak{L}$ to $\mathfrak{L}'$.

3.7.5 Existence of Root Systems, Semisimple Lie Algebras, and Irreducible Modules

We have seen that every irreducible root system is of one of certain types, namely $A_n (n \geq 1)$, $B_n (n \geq 2)$, $C_n (n \geq 3)$, $D_n (n \geq 4)$, $E_n (n = 6, 7, 8)$, $F_n (n = 4)$, $G_n (n = 2)$. Each of these types occurs as the type of some root system. In fact, for any field k of characteristic 0 and for each of these types, there is a split simple Lie algebra over k having a root system of the given type. We do not give the details here. For a complete account, we refer the reader to [13], which also gives the construction of the irreducible modules of a split simple Lie algebra of characteristic 0. We do briefly describe the split simple Lie algebras with root systems of type A_n, B_n, C_n, D_n. The five remaining types E_6, E_7, E_8, F_4, G_2 are called *exceptional*. For their existence, see [28, Part VI, p. 19]. For further description, see [16], [2], [8], [15].

Now we let k be any field of characteristic 0.

The Lie algebra of linear transformations of trace 0 in an $(n + 1)$-

dimensional vector space over k is a split simple Lie algebra over k with root system of type $A_n (n \geq 1)$.

For $n \geq 2$, the Lie algebra of linear transformations in a $(2n + 1)$-dimensional vector space over k which are skew relative to a nondegenerate symmetric bilinear form of maximal Witt index is a split simple Lie algebra over k with a root system of type B_n.

For $n \geq 3$, the Lie algebra of linear transformations in a $2n$-dimensional vector space over k which are skew relative to a nondegenerate skew bilinear form is a split simple Lie algebra over k with root system of type C_n.

For $n \geq 4$, the Lie algebra of linear transformations in a $2n$-dimensional vector space over k which are skew symmetric with respect to a nondegenerate bilinear form of maximal Witt index is a split simple Lie algebra over k with root system of type D_n.

3.8 Automorphisms; Conjugacy Theorems

In this section, we discuss the automorphism group Aut $\mathfrak{L}$ of a Lie algebra $\mathfrak{L}$ over a field k of characteristic 0. We show for k algebraically closed that the Zariski connected component of the identity of Aut $\mathfrak{L}$ acts transitively on the set of Cartan subalgebras of $\mathfrak{L}$ and on the set of maximal solvable subalgebras of $\mathfrak{L}$. This is used to give a reasonably detailed description of Aut $\mathfrak{L}$ for k algebraically closed and $\mathfrak{L}$ semisimple. We then carry over part of the discussion to split semisimple Lie algebras.

We begin by introducing a subgroup $E^{\mathfrak{L}}$ of Aut $\mathfrak{L}$ that is easy to work with because it has nice functorial properties.

3.8.1 Definition

An element x of $\mathfrak{L}$ is *strongly nilpotent* if there exists a $y \in \mathfrak{L}$ such that $x \in \mathfrak{L}_\alpha(\text{ad } y)$ for some $\alpha \neq 0$. The set of strongly nilpotent elements of $\mathfrak{L}$ is denoted $\mathfrak{L}_N$.

If $\mathfrak{B}$ is a subalgebra of $\mathfrak{L}$, then $\mathfrak{B}_N \subset \mathfrak{L}_N$.

For $x \in \mathfrak{L}_N$, $\text{ad}_{\mathfrak{L}} x$ is nilpotent. In showing this, we may assume that k is algebraically closed since $\mathfrak{L}_0(\text{ad}_{\mathfrak{L}}\, x)_K = (\mathfrak{L}_K)_0(\text{ad } \mathfrak{L}_K x)$ for K the algebraic closure of k. But then $\mathfrak{L} = \sum_{\beta \in k} \mathfrak{L}_\beta(\text{ad}_{\mathfrak{L}}\, y)$ with $x \in \mathfrak{L}_\alpha(\text{ad}_{\mathfrak{L}}\, y)$ and $\alpha \neq 0$, for some y. Then $(\text{ad}_{\mathfrak{L}}\, x)^n$ maps $\mathfrak{L}_\beta(\text{ad}_{\mathfrak{L}}\, y)$ into $\mathfrak{L}_{\beta + n\alpha}(\text{ad}_{\mathfrak{L}}\, y)$, and we can choose n such that $\mathfrak{L}_\beta(\mathrm{ad}_\mathfrak{L}\, y)(\mathrm{ad}_\mathfrak{L}\, x)^n = 0$ for all β since $L_\beta(\mathrm{ad}_\mathfrak{L}\, y) = 0$ for all but finitely many β. For such an n, $(\mathrm{ad}_\mathfrak{L}\, x)^n = 0$.

3.8.2 Definition

Let $\mathfrak{B}$ be a subalgebra of $\mathfrak{L}$. Then $\mathrm{Ad}_\mathfrak{L}\, \mathfrak{B}$ is the subgroup of $\mathrm{Aut}\, \mathfrak{L}$ generated by $\{\exp \mathrm{ad}_\mathfrak{L}\, x \mid x \in \mathfrak{B},\ \mathrm{ad}_\mathfrak{L}\, x \text{ is nilpotent}\}$. We let $E_\mathfrak{L}^\mathfrak{B}$ be the subgroup of $\mathrm{Ad}_\mathfrak{L}\, \mathfrak{B}$ generated by $\{\exp \mathrm{ad}_\mathfrak{L}\, x \mid x \in \mathfrak{B}_N\}$. We let $E^\mathfrak{L}$ denote $E_\mathfrak{L}^\mathfrak{L}$.

Obviously, $E_\mathfrak{L}^\mathfrak{B}$ is a subgroup of $E^\mathfrak{L}$ such that $E^\mathfrak{B} = \{g|_\mathfrak{B} \mid g \in E_\mathfrak{L}^\mathfrak{B}\}$.

If $f\colon \mathfrak{L} \to \mathfrak{L}'$ is a surjective homomorphism of Lie algebras and $y \in \mathfrak{L}$, one sees easily that $f(\sum_{\alpha \in k} \mathfrak{L}_\alpha(\mathrm{ad}\, y)) = \sum \mathfrak{L}'_\alpha(\mathrm{ad}\, f(y))$. It follows that $f(\mathfrak{L}_\alpha(\mathrm{ad}\, y)) = \mathfrak{L}'_\alpha(\mathrm{ad}\, f(y))$. Thus, $f(\mathfrak{L}_N) = \mathfrak{L}'_N$. We use this for the following.

3.8.3 Proposition

Let $f\colon \mathfrak{L} \to \mathfrak{L}'$ be a surjective homomorphism of Lie algebras. Then any element g' of $E^{\mathfrak{L}'}$ can be lifted to an element g of $E^\mathfrak{L}$ such that

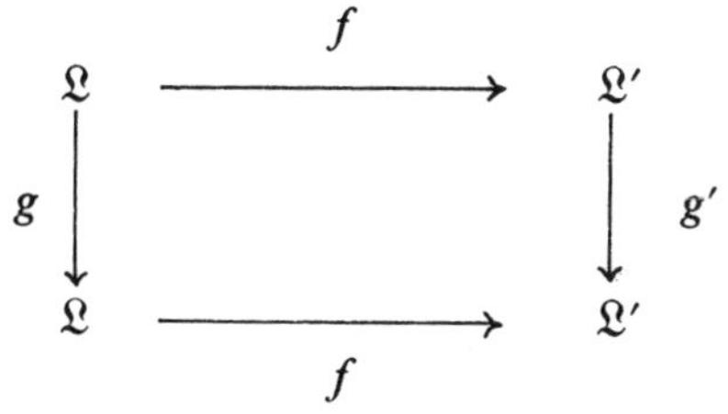

Figure 31.

is commutative.

PROOF. If $x' \in \mathfrak{L}'_N$ and $x \in \mathfrak{L}_N$ such that $f(x) = x'$, then

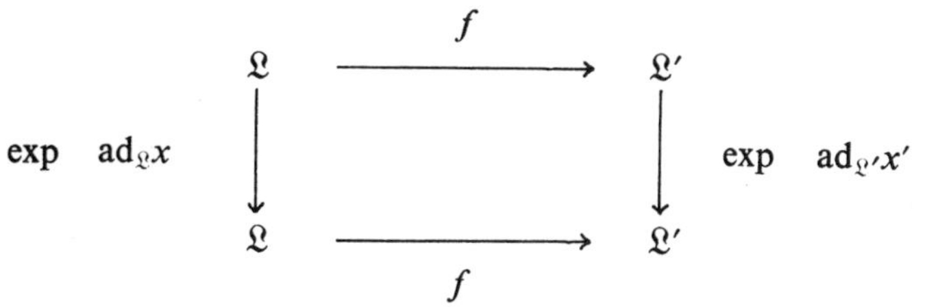

Figure 32.

is commutative. Since every $x' \in \mathfrak{L}'_N$ lifts to an element $x \in \mathfrak{L}_N$ with $f(x) = x'$, and since any $g' \in E^{\mathfrak{L}'}$ is a product of elements of $E^{\mathfrak{L}'}$ of the form $\exp \operatorname{ad}_{\mathfrak{L}'} x'$ with $x' \in \mathfrak{L}'_N$, it follows that any $g' \in E^{\mathfrak{L}'}$ can be lifted to a $g \in E^{\mathfrak{L}}$ such that

$$\begin{array}{ccc} \mathfrak{L} & \xrightarrow{\quad f \quad} & \mathfrak{L}' \\ {\scriptstyle g}\downarrow & & \downarrow{\scriptstyle g'} \\ \mathfrak{L} & \xrightarrow{\quad f \quad} & \mathfrak{L}' \end{array}$$

Figure 33.

is commutative.

3.8.4 Theorem

Let $\mathfrak{L}$ be a split solvable Lie algebra over k. Then $E^{\mathfrak{L}}$ acts transitively on the set of Cartan subalgebras of $\mathfrak{L}$. Consequently, every Cartan subalgebra of $\mathfrak{L}$ is split.

PROOF. The proof is by induction on $\dim \mathfrak{L}$ and is trivial if $\dim \mathfrak{L} = 1$ or $\mathfrak{L}$ is nilpotent. Thus, let $\mathfrak{L}$ be nonnilpotent and let $\mathfrak{H}_1$, $\mathfrak{H}_2$ be Cartan subalgebras of $\mathfrak{L}$ with $\mathfrak{H}_2$ of the form $\mathfrak{H}_2 = \mathfrak{L}_0(\operatorname{ad} x)$ with $x \in \mathfrak{L}_{\text{reg}}$. Let $\mathfrak{A}$ be a minimal nonzero Abelian ideal of $\mathfrak{L}$. (Note that $\mathfrak{L}$ does have nonzero Abelian ideals, e.g., $\mathfrak{L}^{(i)}$ where $\mathfrak{L}^{(i)} \neq \{0\}$ and $\mathfrak{L}^{(i+1)} = \{0\}$). Let $\mathfrak{L}' = \mathfrak{L}/\mathfrak{A}$ and $f: \mathfrak{L} \to \mathfrak{L}'$ be the canonical homomorphism. Let $x' = f(x)$ and $S' = \{x' \mid x \in S\}$ for $x \in \mathfrak{L}$, $S \subset \mathfrak{L}$. Then $\mathfrak{H}'_i$ is a Cartan subalgebra of $\mathfrak{L}'$ since $f(\mathfrak{H}_i) = f(\mathfrak{L}_0(\operatorname{ad} \mathfrak{H}_i)) = \mathfrak{L}'_0(\operatorname{ad} f(\mathfrak{H}_i))$ for $i = 1, 2$ (see the preceding proof). By induction, there exists a $g' \in E^{\mathfrak{L}'}$ such that $\mathfrak{H}_1'^{g'} = \mathfrak{H}'_2$. Lift g' to $g \in E^{\mathfrak{L}}$ such that

$$\begin{array}{ccc} \mathfrak{L} & \longrightarrow & \mathfrak{L}' \\ {\scriptstyle g}\downarrow & & \downarrow{\scriptstyle g'} \\ \mathfrak{L} & \longrightarrow & \mathfrak{L}' \end{array}$$

Figure 34.

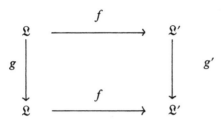

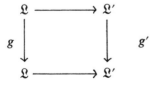

is commutative, and let $\mathfrak{B}_i = f^{-1}(\mathfrak{H}_i')$ for $i = 1, 2$. Then $\mathfrak{B}_1^g = \mathfrak{B}_2$ and $\mathfrak{H}_1^g$, $\mathfrak{H}_2$ are Cartan subalgebras of $\mathfrak{B}_2$. If $\mathfrak{B}_2 \neq \mathfrak{L}$, then there exists an $h \in E_{\mathfrak{L}}^{\mathfrak{B}_2} \subset E^{\mathfrak{L}}$ such that $\mathfrak{H}_1^{gh} = \mathfrak{H}_2$, and we are done. (Recall here that $E^{\mathfrak{B}_2} = \{h|_{\mathfrak{B}_2} \mid h \in E_{\mathfrak{L}}^{\mathfrak{B}_2}\}$). Thus, we may assume that $\mathfrak{B}_2 = \mathfrak{L}$. Then $\mathfrak{B}_1^g = \mathfrak{B}_2 = \mathfrak{L}$ and $\mathfrak{B}_1 = \mathfrak{L}$. Thus, $\mathfrak{L} = \mathfrak{H}_1 + \mathfrak{A} = \mathfrak{H}_2 + \mathfrak{A}$. Let $\mathfrak{B} = \mathfrak{L}_*(\text{ad } x)$, so that $\mathfrak{L} = \mathfrak{H}_2 \oplus \mathfrak{B}$ and $\mathfrak{H}_2\mathfrak{B} \subset \mathfrak{B}$. Since $\mathfrak{A}$ is ad x-stable, $\mathfrak{B} = \mathfrak{A}_*(\text{ad } x) \subset \mathfrak{A}$. Thus, $\mathfrak{B}$ is an Abelian ideal of $\mathfrak{L}$. Thus, $\mathfrak{B} = \mathfrak{A}$ by the minimality of $\mathfrak{A}$, so we have $\mathfrak{L} = \mathfrak{H}_1 + \mathfrak{A} = \mathfrak{H}_1 + \mathfrak{B}$. Let $x = y + b$ with $y \in \mathfrak{H}_1$ and $b \in \mathfrak{B}$. Choose $c \in \mathfrak{B}$ such that $c \text{ ad } x = b$. Then exp ad $c = I + \text{ad } c$ since $\mathfrak{B}$ is an Abelian ideal, and $x \exp \text{ad } c = x + xc = x - b = y$. Since $x \in \mathfrak{L}_{\text{reg}}$, then $y \in \mathfrak{L}_{\text{reg}}$. Thus, $\mathfrak{L}_0(\text{ad } y) = \mathfrak{H}$ is a Cartan subalgebra of $\mathfrak{L}$ containing $\mathfrak{H}_1$. But then $\mathfrak{H} = \mathfrak{H}_1$ since $\mathfrak{H}_1 \subset \mathfrak{H} = \mathfrak{L}_0(\text{ad } \mathfrak{H}) \subset \mathfrak{L}_0(\text{ad } \mathfrak{H}_1) = \mathfrak{H}_1$. Thus, $\mathfrak{H}_1 = \mathfrak{L}_0(\text{ad } y) = \mathfrak{L}_0(\text{ad } (x \exp \text{ad } c)) = \mathfrak{L}_0(\text{ad } x) \exp \text{ad } c = \mathfrak{H}_2 \exp \text{ad } c$. It remains only to show that exp ad $c \in E^{\mathfrak{L}}$. But $\mathfrak{L}$ is split and therefore has a split Cartan subalgebra $\mathfrak{H}$. Since $\mathfrak{H} = \mathfrak{H}_2 \exp \text{ad } d$ for some $d \in \mathfrak{B}$, $\mathfrak{H}_2$ is also split. Now $\mathfrak{B} = \sum_{a \neq 0} \mathfrak{B}_a(\mathfrak{H}_2)$ and $c = \sum c_a$ with $c_a \in \mathfrak{B}_a(\mathfrak{H}_2)$ for all $a \neq 0$. Since the c_a commute, as elements of $\mathfrak{B}$, exp ad $c = \prod \exp \text{ad } c_a$. Since the c_a are in $\mathfrak{L}_N$, exp ad c is therefore in $E^{\mathfrak{L}}$.

If $\mathfrak{L}$ is solvable (not necessarily split), then ad x is nilpotent for $x \in \mathfrak{L}^\infty = \bigcap_{i=1}^\infty \mathfrak{L}^i$, by 3.4.3. The preceding proof can be modified slightly to give the following proposition. The details are left to the reader.

3.8.5 Theorem

Let $\mathfrak{L}$ be solvable. Then $\text{Ad}_{\mathfrak{L}}\, \mathfrak{L}^\infty$ acts transitively on the Cartan subalgebras of $\mathfrak{L}$.

3.8.6 Lemma

Let $\mathfrak{B}$ be a solvable subalgebra of $\mathfrak{L}$, $\mathfrak{B}_0$ a subalgebra of $\mathfrak{B}$ with $\mathfrak{B}_0 \subsetneq \mathfrak{B}$. Let $\mathfrak{N} = \{x \in \mathfrak{B}_0 \mid \text{ad}_{\mathfrak{L}}\, x \text{ is nilpotent}\}$. Then $\mathfrak{N}_{\mathfrak{B}}(\mathfrak{N}) \supsetneq \mathfrak{B}_0$.

PROOF. $\mathfrak{N}$ is an ideal of $\mathfrak{B}_0$, by 3.4.4, so that $\mathfrak{N}_{\mathfrak{B}}(\mathfrak{N}) \supset \mathfrak{B}_0$. For $x \in \mathfrak{N}$, let $f(x)$ be the linear transformation of $\mathfrak{B}/\mathfrak{B}_0$ induced by $\text{ad}_{\mathfrak{B}}\, x$. Then f is a representation of $\mathfrak{N}$ and $f(x)$ is nilpotent for $x \in \mathfrak{N}$. Thus, there exists

$y \in \mathfrak{B} - \mathfrak{B}_0$ such that $(y + \mathfrak{B}_0)f(x) = \mathfrak{B}_0$ for all $x \in \mathfrak{N}$. But then $y\mathfrak{N} \subset \mathfrak{B}_0$. Since $y\mathfrak{N} \subset \mathfrak{B}^{(1)}$ and $\operatorname{ad}_{\mathfrak{L}} \mathfrak{B}^{(1)}$ consists of nilpotent elements, by 3.4.3, $y\mathfrak{N} \subset \mathfrak{N}$ and $y \in \mathfrak{N}_{\mathfrak{B}}(\mathfrak{N})$. Thus, $\mathfrak{N}_{\mathfrak{B}}(\mathfrak{N}) \supsetneqq \mathfrak{B}_0$.

3.8.7 Lemma

Let $\mathfrak{B}$ be a maximal solvable subalgebra of $\mathfrak{L}$. Then $\mathfrak{N}_{\mathfrak{L}}(\mathfrak{B}) = \mathfrak{B}$.

PROOF. Let $x \in \mathfrak{N}_{\mathfrak{L}}(\mathfrak{B})$. Then $kx + \mathfrak{B}$ is a solvable subalgebra of $\mathfrak{L}$, so $kx + \mathfrak{B} = \mathfrak{B}$ and $x \in \mathfrak{B}$.

3.8.8 Theorem

Let k be algebraically closed. Then

1. the Cartan subalgebras of $\mathfrak{L}$ are conjugate under $E^{\mathfrak{L}}$;
2. the maximal solvable subalgebras of $\mathfrak{L}$ are conjugate under $E^{\mathfrak{L}}$.

PROOF. We first note that 2 implies 1. Thus, assume that 2 is true and let $\mathfrak{H}_1$, $\mathfrak{H}_2$ be Cartan subalgebras of $\mathfrak{L}$. Let $\mathfrak{B}_1$, $\mathfrak{B}_2$ be maximal solvable subalgebras of $\mathfrak{L}$ containing $\mathfrak{H}_1$, $\mathfrak{H}_2$ respectively. Take $g \in E^{\mathfrak{L}}$ such that $\mathfrak{B}_2 = \mathfrak{B}_1 g$, by 2. Then there exists $h \in E_{\mathfrak{L}}^{\mathfrak{B}_2} \subset E^{\mathfrak{L}}$ such that $\mathfrak{H}_1 gh = \mathfrak{H}_2$, by 3.8.4. Thus, 2 implies 1.

We now prove 2 by induction on $\dim \mathfrak{L}$. This is trivial if $\dim \mathfrak{L} = 1$ or $\mathfrak{L}$ is solvable. Thus, let $\mathfrak{L}$ be nonsolvable, and let $\mathfrak{R}$ be the radical of $\mathfrak{L}$. Since a subalgebra $\mathfrak{B}$ of $\mathfrak{L}$ is maximal solvable iff $\mathfrak{B} \supset \mathfrak{R}$ and $\mathfrak{B}/\mathfrak{R}$ is a maximal solvable subalgebra of $\mathfrak{L}/\mathfrak{R}$, we may assume that $\mathfrak{R} = \{0\}$. For otherwise, we can apply the induction hypothesis to $\mathfrak{L}/\mathfrak{R}$ and invoke 3.8.3.

Thus, let $\mathfrak{L}$ be semisimple and let X be the set of maximal solvable subalgebras of $\mathfrak{L}$. Let $\mathfrak{T}$ be a Cartan subalgebra of $\mathfrak{L}$, Σ be the root system of $\mathfrak{L}$ with respect to $\mathfrak{T}$, Σ^+ be a half-system of Σ, and $\mathfrak{B} = \mathfrak{T} + \sum_{a \in \Sigma^+} \mathfrak{L}_a$ where $\mathfrak{L}_a = \mathfrak{L}_a(\mathfrak{T})$ for $a \in \Sigma$. We claim that $\mathfrak{B} \in X$. Letting $\mathfrak{N} = \sum_{a \in \Sigma^+} \mathfrak{L}_a$, we have

$$\mathfrak{N}^i \subset \sum_{\substack{a \in \Sigma^+ \\ \mathfrak{L}(a) \geq i}} \mathfrak{L}_a,$$

by a simple induction argument.

Thus, $\mathfrak{N}^i = \{0\}$ for $i = \max_{a \in \Sigma^+} \mathfrak{L}(a) + 1$, and $\mathfrak{N}$ is nilpotent. Thus, $\mathfrak{B}$ is solvable. If $\mathfrak{C} \supsetneqq \mathfrak{B}$, then $\mathfrak{C} = \sum_{\mathfrak{L}_a \subset \mathfrak{C}} \mathfrak{L}_a$, since $\mathfrak{T} \subset \mathfrak{C}$ and, choosing a such that $\mathfrak{L}_a \subset \mathfrak{C}$ and $a \notin \Sigma^+$, $\mathfrak{C}$ contains the simple subalgebra $\mathfrak{L}^{(a)} = \mathfrak{L}_a + \mathfrak{T}_a + \mathfrak{L}_{-a}$ and $\mathfrak{C}$ is not solvable. Thus, $\mathfrak{B}$ is maximal solvable.

We next show that every $\mathfrak{B} \in X$ such that $\mathfrak{B} \cap \mathfrak{B}' \neq \{0\}$ is conjugate to $\mathfrak{B}$ under $E^{\mathfrak{L}}$. Suppose not, and take $\mathfrak{B}' \in X$ with $\mathfrak{B} \cap \mathfrak{B}' \neq \{0\}$ such that $\mathfrak{B}'$ is not conjugate to $\mathfrak{B}$ under $E_{\mathfrak{L}}$ and $\dim \mathfrak{B} \cap \mathfrak{B}'$ is maximal.

There are two cases. In the first, we assume that the ideal $\mathfrak{N} = \{x \in \mathfrak{B} \cap \mathfrak{B}' \mid \operatorname{ad} x$ is nilpotent$\}$ of $\mathfrak{B} \cap \mathfrak{B}'$ is nonzero. Let $\mathfrak{D}$ be the normalizer of $\mathfrak{N}$ in $\mathfrak{L}$. Since $\mathfrak{B} \cap \mathfrak{B}' \subsetneqq \mathfrak{B}$ and $\mathfrak{B} \cap \mathfrak{B}' \subsetneqq \mathfrak{B}'$, we have, by 3.8.6, that $\mathfrak{B} \cap \mathfrak{B}' \subsetneqq \mathfrak{B}_1$ and $\mathfrak{B} \cap \mathfrak{B}' \subsetneqq \mathfrak{B}_1'$ where $\mathfrak{B}_1 = \mathfrak{B} \cap \mathfrak{D}$ and $\mathfrak{B}_1' = \mathfrak{B}' \cap \mathfrak{D}$. Take maximal solvable subalgebras $\mathfrak{B}_2$, $\mathfrak{B}_2'$ of $\mathfrak{D}$ containing $\mathfrak{B}_1$, $\mathfrak{B}_1'$ respectively. Then there exists $g \in E_{\mathfrak{L}}^{\mathfrak{D}} \subset E^{\mathfrak{L}}$ such that $\mathfrak{B}_2 = \mathfrak{B}_2' g$, by the induction hypothesis; for $\mathfrak{D} \neq \mathfrak{L}$, since $\mathfrak{N}$ cannot be an ideal of $\mathfrak{L}$. And there exists $h \in E^{\mathfrak{L}}$ such that $\mathfrak{B} \supset \mathfrak{B}_2' gh$, by the maximality of $\dim (\mathfrak{B} \cap \mathfrak{B}')$, since $\mathfrak{B} \cap \mathfrak{B}_2' g = \mathfrak{B} \cap \mathfrak{B}_2 \supset \mathfrak{B}_1 \supsetneqq \mathfrak{B} \cap \mathfrak{B}'$. But now $\mathfrak{B} \cap \mathfrak{B}' gh \supset \mathfrak{B}_2' gh \cap \mathfrak{B}' gh \supset \mathfrak{B}_1' gh \supsetneqq (\mathfrak{B} \cap \mathfrak{B}') gh$ so that $\dim (\mathfrak{B} \cap \mathfrak{B}' gh) > \dim \mathfrak{B} \cap \mathfrak{B}'$. Thus, there exists an $f \in E^{\mathfrak{L}}$ such that $\mathfrak{B} = \mathfrak{B}' ghf$, a contradiction.

In the remaining case, the above $\mathfrak{N}$ is $\{0\}$ and $\mathfrak{B} \cap \mathfrak{B}' = \mathfrak{S}$ is nonzero. We first note that $\mathfrak{S}$ is a torus (see 3.7.2.1). Since $\mathfrak{S}^{(1)} \subset \mathfrak{N} = \{0\}$, $\mathfrak{S}$ is Abelian. Now let $x \in \mathfrak{S} - \{0\}$. Then $x = x_s + x_n$, and $x_s \neq 0$ since $\mathfrak{N} = \{0\}$. Now x normalizes $\mathfrak{B}$ and $\mathfrak{B}'$, so x_n does also, by 1.4.7. Thus, $x_n \in \mathfrak{B}$ and $x_n \in \mathfrak{B}'$, by 3.8.7, so $x_n \in \mathfrak{N} = \{0\}$ and $x_n = 0$. Thus, $x = x_s$ and x is semisimple. Thus, $\mathfrak{S}$ is a torus.

We next note that $\mathfrak{S}$ is contained in a Cartan subalgebra $\mathfrak{S}_1$ of $\mathfrak{B}$. For this, let $\mathfrak{C} = \mathfrak{B}_0(\operatorname{ad} \mathfrak{S})$ and let $\mathfrak{S}_1$ be a Cartan subalgebra of $\mathfrak{C}$. Since $\mathfrak{C}\mathfrak{S} = 0$ and $\mathfrak{S} \subset \mathfrak{C}$, we have $\mathfrak{S} \subset \mathfrak{N}_{\mathfrak{C}}(\mathfrak{S}_1) = \mathfrak{S}_1$. Now $\mathfrak{B}_0(\operatorname{ad} \mathfrak{S}_1) \subset \mathfrak{B}_0(\operatorname{ad} \mathfrak{S}) = \mathfrak{C}$ and we have $\mathfrak{S}_1 \subset \mathfrak{B}_0(\operatorname{ad} \mathfrak{S}_1) = \mathfrak{C}_0(\operatorname{ad} \mathfrak{S}_1) = \mathfrak{S}_1$. Thus, $\mathfrak{S}_1 = \mathfrak{B}_0(\operatorname{ad} \mathfrak{S}_1)$ and $\mathfrak{S}_1$ is a Cartan subalgebra of $\mathfrak{B}$ containing $\mathfrak{S}$.

Since $\mathfrak{S}_1$ and $\mathfrak{T}$ are conjugate under $E_{\mathfrak{L}}^{\mathfrak{B}} \subset E^{\mathfrak{L}}$, by 3.8.4, we may assume that $\mathfrak{S}_1 = \mathfrak{T}$, that is, that $\mathfrak{S} \subset \mathfrak{T}$. If $\mathfrak{S} = \mathfrak{T}$, then $\mathfrak{T} \subset \mathfrak{B}'$ and $\mathfrak{B}' = \sum_{\mathfrak{L}_a \subset \mathfrak{B}'} \mathfrak{L}_a$, so that $\mathfrak{B}' \supset \mathfrak{L}_a$ with $-a \in \Sigma^+$ for some a. In this case, $\mathfrak{B} \cap (\mathfrak{B}' w_a) \supset \mathfrak{T} + \mathfrak{L}_{-a}$ and $\dim (\mathfrak{B} \cap \mathfrak{B}' w_a) > \dim \mathfrak{B} \cap \mathfrak{B}'$ where w_a is the element of the Weyl group of $\mathfrak{L}$ described in 3.7.2.4. But then $\mathfrak{B} = \mathfrak{B}' w_a g$ for some $g \in E^{\mathfrak{L}}$. Since $w_a \in E^{\mathfrak{L}}$, this is a contradiction. Thus, $\mathfrak{S} \subsetneqq \mathfrak{T}$.

Consider the decomposition $\mathfrak{B}' = \sum \mathfrak{B}'_a(\mathfrak{S})$ and choose a, b and $s \in \mathfrak{S} - \{0\}$ such that $a \in \mathfrak{S}^*$, $b \in \mathfrak{B}'_a(\mathfrak{S}) - \mathfrak{S}$, and $a(s)$ is a nonnegative rational number. This is possible, for otherwise $\mathfrak{B}' = \mathfrak{S}$ and $\mathfrak{B}'$ is not maxima solvable. If $a(s) > 0$, let $\mathfrak{B}'' = \mathfrak{T} + \sum_{b(s)>0} \mathfrak{L}_b$ where "$b(s) > 0$" is the statement "$b(s)$ is a positive rational number." Then $\mathfrak{B}''$ is solvable, by an argument similar to the one showing that $\mathfrak{B} = \mathfrak{T} + \sum_{a\in\Sigma^+} \mathfrak{L}_a$ is solvable. And $\mathfrak{B} \cap \mathfrak{B}'' \supset \mathfrak{T} \supsetneq \mathfrak{B} \cap \mathfrak{B}'$, so that there exists $g \in E^{\mathfrak{L}}$ such that $\mathfrak{B} \supset \mathfrak{B}''g$, by the minimality of $\mathfrak{B} \cap \mathfrak{B}'$. But then $\mathfrak{B} \cap \mathfrak{B}'g \supset \mathfrak{B}''g \cap \mathfrak{B}'g \supset (\mathfrak{S} + kb)g$ and $\dim \mathfrak{B} \cap \mathfrak{B}'g > \dim \mathfrak{B} \cap \mathfrak{B}'$. Thus, there exists $h \in E^{\mathfrak{L}}$ such that $\mathfrak{B} = \mathfrak{B}'gh$, by the maximality of $\mathfrak{B} \cap \mathfrak{B}'$, a contradiction. If $a(s) = 0$, let $\mathfrak{B}'' = \mathfrak{S} \oplus kb$ and note that $\mathfrak{B}''$ is a solvable subalgebra of $\mathfrak{C}_{\mathfrak{L}}(s)$. Now $\mathfrak{C}_{\mathfrak{L}}(s) = \mathfrak{T} + \sum_{a(s)=0} \mathfrak{L}_a$ and

$$\mathfrak{B} \cap \mathfrak{C}_{\mathfrak{L}}(s) = \mathfrak{T} + \sum_{\substack{a(s)=0\\ a\in\Sigma^+}} \mathfrak{L}_a$$

is a maximal solvable subalgebra of $\mathfrak{C}_{\mathfrak{Z}}(s)$. (The proof is the same as that showing that $\mathfrak{B} = \mathfrak{T} + \sum_{a\in\Sigma^+} \mathfrak{L}_a$ is maximal solvable in $\mathfrak{L}$.) Since $\dim \mathfrak{C}_{\mathfrak{L}}(s) < \dim \mathfrak{L}$, there exists a $g \in E_{\mathfrak{L}}^{\mathfrak{C}_{\mathfrak{L}}(s)} \subset E^{\mathfrak{L}}$ such that $\mathfrak{B} \cap \mathfrak{C}_{\mathfrak{L}}(s) \supset \mathfrak{B}''g$, by induction. But then $\mathfrak{B} \cap \mathfrak{B}'g \supset (\mathfrak{S} \oplus kb)g$ and $\dim \mathfrak{B} \cap \mathfrak{B}'g > \dim \mathfrak{S} = \dim \mathfrak{B} \cap \mathfrak{B}'$. Thus, there exists an $h \in E^{\mathfrak{L}}$ such that $\mathfrak{B} = \mathfrak{B}'gh$, a contradiction. We have therefore shown that $\mathfrak{B} \cap \mathfrak{B}' \neq \{0\}$ with $\mathfrak{B}' \in X$ implies that $\mathfrak{B}$ and $\mathfrak{B}'$ are conjugate under $E^{\mathfrak{L}}$.

It remains only to show that $\mathfrak{B} \cap \mathfrak{B}' \neq \{0\}$ for every $\mathfrak{B}' \in X$. Thus, let $\mathfrak{B}' \in X$ and let $\mathfrak{S}'$ be a maximal torus of $\mathfrak{B}'$. Then $\mathfrak{S}' \neq \{0\}$, by 3.8.6 and 3.8.7, for otherwise $\mathfrak{B}'$ consists of nilpotent elements and $\mathfrak{N}_{\mathfrak{L}}\mathfrak{B}' \neq \mathfrak{B}'$. Let $\mathfrak{S}$ be a maximal torus of $\mathfrak{L}$ containing $\mathfrak{S}'$. Then $\mathfrak{S}$ is a Cartan subalgebra of $\mathfrak{L}$, by 3.7.2.1, so that $\mathfrak{B}'' = \mathfrak{S} + \sum_{a\in\Sigma''^+} \mathfrak{L}_a(\mathfrak{S})$ is in X where Σ''^+ is a half-system of the root system Σ'' of $\mathfrak{L}$ with respect to $\mathfrak{S}$. Now $2\dim \mathfrak{B}'' - \dim \mathfrak{S} = 2\dim \mathfrak{B} - \dim \mathfrak{T} = \dim \mathfrak{L}$, so that $2\dim \mathfrak{B}'' > \dim \mathfrak{L}$ and $2\dim \mathfrak{B} > \dim \mathfrak{L}$. Thus, $\mathfrak{B} \cap \mathfrak{B}'' \neq 0$. Thus, $\mathfrak{B} = \mathfrak{B}''g$ for some $g \in E^{\mathfrak{L}}$. But then $\mathfrak{B} \cap \mathfrak{B}'g \supset \mathfrak{S}g \neq 0$ and $\mathfrak{B} = \mathfrak{B}'gh$ for some $h \in E^{\mathfrak{L}}$. Thus, $2\dim \mathfrak{B}' = 2\dim \mathfrak{B} > \dim \mathfrak{L}$. It follows that $\mathfrak{B} \cap \mathfrak{B}' \neq \{0\}$.

3.8.9 Corollary

Let $\mathfrak{L}$ be a Lie algebra over k, k algebraically closed. Then

1. every maximal solvable subalgebra of $\mathfrak{L}$ contains a Cartan subalgebra of $\mathfrak{L}$;
2. $E^{\mathfrak{L}}$ acts transitively on the pairs $(\mathfrak{B}, \mathfrak{H})$ where $\mathfrak{B}$ is a maximal solvable subalgebra of $\mathfrak{L}$ and $\mathfrak{H}$ a Cartan subalgebra of $\mathfrak{B}$;
3. if $\mathfrak{L}$ is semisimple with Cartan subalgebra $\mathfrak{T}$, root system Σ with respect to $\mathfrak{T}$, and simple system π of Σ, then any $f \in \text{Aut}\,\mathfrak{L}$ can be written $f = gh$ where $g \in E^{\mathfrak{L}}$, $\mathfrak{T}h = \mathfrak{T}$, and $\pi h^* = \pi$ (see 3.7.2.6). Moreover, $\{h^* \mid h \in \text{Aut}\,\mathfrak{L},\ \mathfrak{T}h = \mathfrak{T},\ \pi h^* = \pi\}$ is the finite group $\text{Aut}\,(\Sigma, \pi)$ of all automorphisms of Σ which stabilize π.

PROOF. If $\mathfrak{H}$ is a Cartan subalgebra of $\mathfrak{L}$, then $\mathfrak{H}$ is contained in a maximal solvable subalgebra $\mathfrak{B}$ of $\mathfrak{L}$. Since any maximal solvable subalgebra $\mathfrak{B}'$ of $\mathfrak{L}$ is conjugate to $\mathfrak{B}$ under $E^{\mathfrak{L}}$, such a $\mathfrak{B}'$ must contain a Cartan subalgebra of $\mathfrak{L}$ since $\mathfrak{B}$ does. This proves 1.

For 2, let $\mathfrak{B}$, $\mathfrak{B}'$ be maximal solvable subalgebras of $\mathfrak{L}$, and let $\mathfrak{H}$, $\mathfrak{H}'$ be Cartan subalgebras of $\mathfrak{B}$, $\mathfrak{B}'$. Choose $g \in E^{\mathfrak{L}}$ such that $\mathfrak{B} = \mathfrak{B}'g$, and then choose $h \in E^{\mathfrak{B}}_{\mathfrak{L}}$ such that $\mathfrak{H} = \mathfrak{H}'gh$, by 3.8.4. Then $(\mathfrak{B}, \mathfrak{H}) = (\mathfrak{B}', \mathfrak{H}')gh$.

For 3, let $\mathfrak{B} = \mathfrak{T} + \sum_{a \in \Sigma^+} \mathfrak{L}_a$ where Σ^+ is the half-system of Σ corresponding to π. Choose $g \in E^{\mathfrak{L}}$ such that $(\mathfrak{B}, \mathfrak{T})g^{-1} = (\mathfrak{B}, \mathfrak{T})f^{-1}$ and let $h = g^{-1}f$. Then $f = gh$ and $(\mathfrak{B}, \mathfrak{T})h = (\mathfrak{B}, \mathfrak{T})$, that is, $\mathfrak{B}h = \mathfrak{B}$ and $\mathfrak{T}h = \mathfrak{T}$. Thus, $\sum_{a \in \Sigma^+} \mathfrak{L}_a = (\sum_{a \in \Sigma^+} \mathfrak{L}_a)h = \sum_{a \in \Sigma^+} (\mathfrak{L}_a h) = \sum_{a \in \Sigma^+} \mathfrak{L}_{ah^*}$. Thus, $\Sigma^+ h^* = \Sigma^+$ so that $\pi h^* = \pi$. That $h^* \in \text{Aut}\,\Sigma$ is obvious. That every automorphism of Σ preserving π occurs as an h^* follows from 3.7.4.9.

One can show that the decomposition $f = gh$ given in 3 above is essentially unique and that the h of this decomposition is in $E^{\mathfrak{L}}$ iff h^* is the identity on Σ. It then follows that $\text{Aut}\,\mathfrak{L}/E^{\mathfrak{L}} = \text{Aut}\,(\Sigma, \pi)$, so that $\text{Aut}\,\mathfrak{L}/E^{\mathfrak{L}}$ is isomorphic to the group $\text{Aut}\,D_\pi$ of automorphisms of the Dynkin diagram of $\mathfrak{L}$ with respect to $\mathfrak{T}$. Details are given in [13], [26]. The complete proof is somewhat lengthy and we do not include it, but the flavor of it is roughly that of the following incomplete proof. In the Zariski topology for $\text{Aut}\,\mathfrak{L}$, $E^{\mathfrak{L}}$ is connected since the generators lie in connected one-parameter groups $\{\exp \text{ad}\, tx \mid t \in k\}$, $\text{ad}\, x$ nilpotent. Thus, $E^{\mathfrak{L}} \subset (\text{Aut}\,\mathfrak{L})_0$ where $(\text{Aut}\,\mathfrak{L})_0$ is the connected component of 1 in $\text{Aut}\,\mathfrak{L}$. Now $\text{Aut}\,\mathfrak{L}$ can be made to act continuously on π, so that $(\text{Aut}\,\mathfrak{L})_0$ acts trivially.

Translated, this means that h^* is the identity on π, hence on Σ, for $h \in (\text{Aut } \mathfrak{L})_0$. And if h^* is the identity on Σ, then $\mathfrak{L}_a h = \mathfrak{L}_a$ and $th = t$ for $a \in \Sigma$ and $t \in \mathfrak{T}$. By 3.7.4.9, the set of such h is connected and contains 1, so that $h \in (\text{Aut } \mathfrak{L})_0$. Thus, $\text{Aut } \mathfrak{L}/(\text{Aut } \mathfrak{L})_0 \simeq \text{Aut }(\Sigma, \pi)$. Since $E^{\mathfrak{L}} \subset (\text{Aut } \mathfrak{L})_0$, the decomposition given in 3 now implies that $E^{\mathfrak{L}} = (\text{Aut } \mathfrak{L})_0$.

The foregoing material has a formulation which can be proved for split Lie algebras over nonalgebraically closed fields. We content ourselves with the following weak version of part of this material.

3.8.10 Theorem

Let $\mathfrak{L}$ be a split semisimple Lie algebra over k. Then
1. the split Cartan subalgebras of $\mathfrak{L}$ are conjugate under Aut $\mathfrak{L}$;
2. the maximal solvable subalgebras of $\mathfrak{L}$ containing a given split Cartan subalgebra $\mathfrak{T}$ of $\mathfrak{L}$ are conjugate under the Weyl group W of $\mathfrak{L}$ with respect to $\mathfrak{T}$.

PROOF. For 1, let K be the algebraic closure of k and let $\mathfrak{T}_1$, $\mathfrak{T}_2$ be split Cartan subalgebras of $\mathfrak{L}$. Then $\mathfrak{T}_{1K}$, $\mathfrak{T}_{2K}$ are Cartan subalgebras of $\mathfrak{L}_K$ and there exists $g \in E^{\mathfrak{L}_K}$ such that $\mathfrak{T}_{1K}g = \mathfrak{T}_{2K}$. Let Σ_i be the root system of $\mathfrak{L}$ with respect to $\mathfrak{T}_i$ $(i = 1, 2)$ and π_1 a simple system of Σ_1. Since

$$\mathfrak{L} = \mathfrak{T}_i + \sum_{a \in \Sigma_i} \mathfrak{L}_a(\mathfrak{T}_i),$$

then

$$\mathfrak{L}_K = \mathfrak{T}_K + \sum (\mathfrak{L}_K)_{a_K}(\mathfrak{T}_{iK})$$

and

$$(\mathfrak{L}_a(\mathfrak{T}_i))_K = (\mathfrak{L}_K)_{a_K}(\mathfrak{T}_{iK}),$$

where $a_K = a \otimes I_K$ for $a \in \Sigma_i$ $(i = 1, 2)$. Thus, $(\mathfrak{L}_a(\mathfrak{T}_1))_K g = (\mathfrak{L}_{a'}(\mathfrak{T}_2))_K$ where $a'_K(tg) = a_K(t)$ for $t \in \mathfrak{T}_{1K}$. The mapping $a \mapsto a'$ is an isomorphism from Σ_1 to Σ_2 and maps π_1 to a simple system π_2 of Σ_2. Let $\pi_1 = \{a_1, \ldots, a_n\}$ and choose nonzero $e_j \in \mathfrak{L}_{a_j}$, $e_{j'} \in \mathfrak{L}_{a_j'}$ for $1 \leq j \leq n$. Upon multiplying g by a suitable automorphism of $\mathfrak{L}_{2K}$ stabilizing $\mathfrak{T}_{2K}$ and each $\mathfrak{L}_{a_j'}$, using 3.7.4.9, we may assume that $e_jg = e_j'$ for $1 \leq j \leq n$. We now show that $\mathfrak{L}g = \mathfrak{L}$, or that $g|_{\mathfrak{L}} \in \text{Aut } \mathfrak{L}$. Choose nonzero $f_j \in \mathfrak{L}_{-a_j}$ and $f_j' \in \mathfrak{L}_{-a_j'}$ and let $h_j = e_jf_j$, $h_j' = e_j'f_j'$. Then $f_jg = \gamma_j f_j'$ with $\gamma_j \in K$ $(1 \leq j \leq n)$. Now $e_jh_j = a_j(h_j)e_j \neq 0$ and we have

$$\begin{aligned} a_j(h_j)e_j' &= a_j(h_j)e_jg = (e_jh_j)g \\ &= (e_j(e_jf_j))g = \gamma_j(e_j'(e_j'f_j')). \end{aligned}$$

Since $a_j(h_j)e_j'$ is a nonzero element of $\mathfrak{L}$ and $e_j'(e_j'f_j') \in \mathfrak{L}$, each γ_j is in k for $1 \leq j \leq n$. Thus, $(\mathfrak{L}_{\pm a_j})g \subset \mathfrak{L}_{\pm a_j}$ for $1 \leq j \leq n$ and g maps a generating set of $\mathfrak{L}$ into a generating set of $\mathfrak{L}$. Thus, $\mathfrak{L}g = \mathfrak{L}$, $\mathfrak{T}_1 g = \mathfrak{T}_2$ and $g|_{\mathfrak{L}} \in \text{Aut } \mathfrak{L}$. This proves 1.

For 2, let Σ be the root system of $\mathfrak{L}$ with respect to $\mathfrak{T}$ and let $\mathfrak{B}_{\Sigma^+} = \mathfrak{T} + \sum_{a\in\Sigma^+} \mathfrak{L}_a$ where Σ^+ is any half-system of Σ. As in the proof of 3.8.8, each such $\mathfrak{B}_{\Sigma^+}$ is maximal solvable. Any two $\mathfrak{B}_{\Sigma_1^+}$, $\mathfrak{B}_{\Sigma_2^+}$ are conjugate under the Weyl group W of $\mathfrak{L}$ with respect to $\mathfrak{T}$, where Σ_1^+, Σ_2^+ are half-systems of Σ, by 3.7.2.7 and 3.7.3.4.2. Thus, it suffices to show that every maximal solvable subalgebra $\mathfrak{B}$ of $\mathfrak{L}$ containing $\mathfrak{T}$ is of the form $\mathfrak{B}_{\Sigma^+}$ for some half-system Σ^+ of Σ. Take such a $\mathfrak{B}$. Then $\mathfrak{B}_K$ is contained in a maximal solvable subalgebra of $\mathfrak{L}_K$ containing $\mathfrak{T}_K$. Thus, it suffices to show that a maximal solvable subalgebra of $\mathfrak{L}_K$ containing $\mathfrak{T}_K$ has the form $\mathfrak{T}_K + \sum_{a\in\Sigma^+}(\mathfrak{L}_a)_K$ for some half-system Σ^+ of Σ, since then $\mathfrak{B} \subset \mathfrak{B}_{\Sigma^+}$, and $\mathfrak{B} = \mathfrak{B}_{\Sigma^+}$ by the maximality of $\mathfrak{B}$. That is, we may assume without loss of generality that k is algebraically closed. But then it follows from 3.8.8 that any maximal solvable subalgebra $\mathfrak{B}'$ of $\mathfrak{L}$ containing $\mathfrak{T}$ is conjugate to $\mathfrak{B}_{\Sigma^+}$ for some half-system Σ^+ of Σ. If $\mathfrak{B}' = \mathfrak{B}_{\Sigma^+}g$, then $\mathfrak{T}g \subset \mathfrak{B}'$ and $\mathfrak{T}gh = \mathfrak{T}$ for some $h \in E^{\mathfrak{B}'}$. Now $\mathfrak{B}_\Sigma gh = \mathfrak{B}'$ and $\mathfrak{T}gh = \mathfrak{T}$, so that $\mathfrak{B}' = \mathfrak{B}_{\Sigma'^+}$ for $\Sigma'^+ = \Sigma^+(gh)^*$, a half-system of Σ.

Lie Algebras of Arbitrary Characteristic

4.1 Introduction

We now turn to Lie algebras of arbitrary characteristic. These Lie algebras are considerably more pathological than those of characteristic 0, and the classification problem is much more difficult. There are simple finite-dimensional Lie algebras of characteristic p that are quite unrelated to any of characteristic 0. The first successful approach in the direction of classifying the simple Lie algebras of characteristic p was that of picking out conditions under which a simple Lie algebra of characteristic p could be shown to be closely related to a simple Lie algebra of characteristic 0. A quite satisfactory account of this is given in [26].

We do not consider the classification problem further in this book, and refer the reader to [19] and [26]. The objective of the present chapter is rather to present general material on Lie algebras of arbitrary characteristic which might lead in the direction of a definitive theory. We are particularly concerned with the structure of Lie algebras in terms of their Cartan subalgebras and other important nilpotent subalgebras.

We continue to follow the language and conventions of Chapter 3. Throughout this chapter, k is a field of characteristic 0 or p, p being a prime. Lie algebras and Lie modules are finite dimensional over k.

We begin by introducing the notions of Lie p-algebra and graded Lie algebra. The importance of Lie p-algebras is that they occur frequently in the theory of groups and of nonassociative algebras, they are less pathological than arbitrary Lie algebras of characteristic p, and finally, any Lie algebra of characteristic p has a nice imbedding in a Lie p-algebra. Graded Lie algebras are important since derivations, automorphisms, and nilpotent subalgebras of a Lie algebra $\mathfrak{L}$ can be used to give $\mathfrak{L}$ the structure of a graded Lie algebra.

4.1.1 Definition

A *Lie p-algebra* (*restricted Lie algebra of characteristic p*) $\mathfrak{L}$ is a Lie algebra over a field k of characteristic p together with a mapping $x \mapsto x^p$ from $\mathfrak{L}$ into $\mathfrak{L}$ such that

1. $(\gamma x)^p = \gamma^p x^p$ for $\gamma \in k$, $x \in \mathfrak{L}$;
2. $(\operatorname{ad} x)^p = \operatorname{ad} x^p$ for $x \in L$;
3. $(x + y)^p = x^p + y^p + \sum_0^{p-1} s_i(x, y)$;

where $is_i(x, y)$ is the coefficient of τ^{i-1} in $x \operatorname{ad} (\tau x + y)^{p-1}$. The mapping $x \mapsto x^p$ is called the *pth power mapping.*

For Lie algebras $\mathfrak{L}$ with trivial center, condition 3 can be dropped. A Lie algebra $\mathfrak{L}$ of characteristic p can be given the structure of a Lie p-algebra iff $\operatorname{ad} x \in \mathfrak{L}$ implies $(\operatorname{ad} x)^p \in \mathfrak{L}$ for all $x \in \mathfrak{L}$. We use this here only indirectly, and refer the reader to [13] for the proof. Using condition 3 (or the preceding observation, e.g., by way of 4.1.4) one shows that if $\mathfrak{L}$ is a Lie p-algebra, there is a unique extension of $x \mapsto x^p$ from $\mathfrak{L}$ to $\mathfrak{L}_{k'}$, where k' is any extension field of k, such that $\mathfrak{L}_{k'}$ together with the extension of $x \mapsto x^p$ is a Lie p-algebra over k'.

The Lie algebra $(\operatorname{Hom} V)_{\text{Lie}}$, with V a finite-dimensional vector space of characteristic p, together with $x \mapsto x^p$ (ordinary pth power of x in $(\operatorname{Hom} V)_{\text{Assoc}}$) is a Lie p-algebra (see [13]). This leads naturally to the notion of linear Lie p-algebra, defined as follows.

4.1.2 Definition

A *p-subalgebra* (respectively *p-ideal*) of a Lie p-algebra $\mathfrak{L}$ is a subalgebra (respectively ideal) of $\mathfrak{L}$ which is stable under $x \mapsto x^p$. Such a subalgebra (respectively ideal) is regarded as a Lie p-algebra by taking its pth power mapping to be the restriction of that of $\mathfrak{L}$.

4.1.3 Definition

A *linear Lie p-algebra* is a p-subalgebra of $(\operatorname{Hom} V)_{\text{Lie}}$, V being a finite-dimensional vector space of characteristic p.

4.1.4 Example

Let $\mathfrak{L}$ be a Lie algebra such that $\operatorname{ad} x \in \operatorname{ad} \mathfrak{L}$ implies that $(\operatorname{ad} x)^p \in \operatorname{ad} \mathfrak{L}$. Then $\operatorname{ad} \mathfrak{L}$ is a linear Lie p-algebra. It then follows easily from condition 3 that $\operatorname{ad} \mathfrak{L}_{k'}$ is a linear Lie p-algebra, that is, that $\operatorname{ad} x \in \operatorname{ad} \mathfrak{L}_{k'}$ implies that $(\operatorname{ad} x)^p \in \operatorname{ad} \mathfrak{L}_{k'}$ for all $\operatorname{ad} x \in \operatorname{ad} \mathfrak{L}_{k'}$.

4.1.5 Example

Let $\mathfrak{A}$ be a finite-dimensional nonassociative algebra of characteristic p. Then $\operatorname{Der} \mathfrak{A}$ is a linear Lie p-algebra, by 2.4.11.

4.1.6 Definition

A *p-homomorphism* (respectively *p-isomorphism*) is a Lie algebra homomorphism (respectively isomorphism) f of Lie p-algebras such that $f(x^p) = f(x)^p$ for $x \in \text{Domain}\, f$.

4.1.7 Definition

If $\mathfrak{J}$ is a p-ideal of a Lie p-algebra $\mathfrak{L}$, the *p-quotient* $\mathfrak{L}/\mathfrak{J}$ of $\mathfrak{L}$ by $\mathfrak{J}$ is the Lie algebra quotient $\mathfrak{L}/\mathfrak{J}$ together with the mapping $x + \mathfrak{J} \mapsto x^p + \mathfrak{J}$ (which is well defined, by condition 3).

One verifies easily that the p-quotient $\mathfrak{L}/\mathfrak{J}$ of a Lie p-algebra $\mathfrak{L}$ by a p-ideal $\mathfrak{J}$ is a Lie p-algebra and that the canonical homomorphism $\mathfrak{L} \to \mathfrak{L}/\mathfrak{J}$ is a p-homomorphism.

The kernel of a p-homomorphism is a p-ideal. The fundamental homomorphism theorems for Lie p-algebras are immediate consequences of those for Lie algebras.

4.1.8 Definition

A *p-representation* of a Lie p-algebra $\mathfrak{L}$ is a Lie algebra representation f of $\mathfrak{L}$ such that $f(x^p) = f(x)^p$ for all $x \in \mathfrak{L}$. A Lie *p-module* for a Lie p-algebra $\mathfrak{L}$ is a Lie module $\mathfrak{M}$ for $\mathfrak{L}$ such that $mx^p = (\ldots(mx)\ldots)x$ (p times) for $m \in \mathfrak{M}$, $x \in \mathfrak{L}$.

Obviously, a faithful Lie module $\mathfrak{M}$ for a Lie p-algebra $\mathfrak{L}$ is a Lie p-module for $\mathfrak{L}$ iff the representation f afforded by $\mathfrak{M}$ is a p-representation of $\mathfrak{L}$.

4.1.9 Proposition

Let $\mathfrak{L}$ be a Lie p-algebra. Then ad is a p-representation of $\mathfrak{L}$.

PROOF. $(\text{ad}\, x)^p = \text{ad}\, x^p$, by condition 2.

4.1.10 Corollary

The center of a Lie p-algebra is a p-ideal.

PROOF. The center is the kernel of ad, and ad is a p-homomorphism.

Throughout the remainder of the section, A and B are groups.

4.1.11 Definition

An *A-graded Lie algebra* (respectively *Lie p-algebra*) is a Lie algebra (respectively Lie p-algebra) $\mathfrak{L}$ together with a decomposition $\mathfrak{L} = \sum_{a \in A} \oplus \mathfrak{L}_a$ (direct sum of subspaces) of $\mathfrak{L}$ such that $\mathfrak{L}_a \mathfrak{L}_b \subset \mathfrak{L}_{ab}$ for all $a, b \in A$.

4.1.12 Example

Let $\mathfrak{L}$ be a Lie algebra over k, $\mathfrak{N}$ a nilpotent subalgebra of Der $\mathfrak{L}$ such that x is split over k for $x \in \mathfrak{N}$. Then $\mathfrak{L} = \sum_{a \in A} \oplus \mathfrak{L}_a(\mathfrak{N})$, A being the additive group of k-valued functions on $\mathfrak{N}$, equips $\mathfrak{L}$ with the structure of an A-graded Lie algebra, by 2.4.12. This shows in particular that if $D \in \operatorname{Der} \mathfrak{L}$, $x \in \mathfrak{L}$ or $\mathfrak{H}$ is a nilpotent subalgebra of $\mathfrak{L}$, and if D, ad x, or ad y ($y \in \mathfrak{H}$) is split over k, then $\mathfrak{L} = \sum_{a \in A} \mathfrak{L}_a(D)$, $\mathfrak{L} = \sum_{a \in A} \mathfrak{L}_a(\operatorname{ad} x)$, or $\mathfrak{L} = \sum_{a \in A} \mathfrak{L}_a(\operatorname{ad} \mathfrak{H})$ equips $\mathfrak{L}$ with the structure of an A-graded Lie algebra. Here, A is the additive subgroup of k generated by the eigenvalues of D, the additive subgroup of k generated by the eigenvalues of ad x, or the additive group of k-valued functions on ad $\mathfrak{H}$.

4.1.13 Example

Let $\mathfrak{L}$ be a Lie algebra, G an Abelian group of automorphisms of $\mathfrak{L}$ such that g is split over k for $g \in G$. Then the decomposition $\mathfrak{L} = \sum_{a \in A} \mathfrak{L}_a(G)$, A being the group of k-valued characters of G, equips $\mathfrak{L}$ with the structure of an A-graded Lie algebra. For this, it suffices to show that for $g \in G$, $x \in \mathfrak{L}_a(g)$, and $y \in \mathfrak{L}_b(g)$, then $xy \in \mathfrak{L}_{ab}(g)$. This is done by showing inductively that $xy(g - abI)^n$ is a sum of terms of the form $u(g - aI)^i v(g - bI)^j$ where $u \in \mathfrak{L}_a(g)$, $v \in \mathfrak{L}_b(g)$ and $i + j = n$. The induction step is proved by noting the following identities for $s = u(g - aI)^i$ and $t = v(g - bI)^j$:

$$\begin{aligned} st(g - abI) &= (sg)(tg) - (sa)(tb) \\ &= s(g - aI)(tg) + (sa)t(g - bI) \\ &= u(g - aI)^{i+1}(vg)(g - bI)^j + (ua)(g - aI)^i v(g - bI)^{j+1}. \end{aligned}$$

Now it is clear that $\mathfrak{L} = \sum_{a \in B} \mathfrak{L}_a(g)$, B being the subgroup of $k - \{0\}$ generated by the eigenvalues of g, equips $\mathfrak{L}$ with the structure of a B-graded

Lie algebra. The more general assertion about $\mathfrak{L} = \sum_{a\in A} \mathfrak{L}_a(G)$ follows directly.

4.1.14 Definition

An *A-graded subalgebra* (respectively *p-subalgebra*) of an A-graded Lie algebra (respectively Lie p-algebra) $\mathfrak{L}$ is a subalgebra (respectively p-subalgebra) $\mathfrak{B}$ of $\mathfrak{L}$ such that $\mathfrak{B} = \sum_{a\in A} \mathfrak{B}_a$ where $\mathfrak{B}_a = \mathfrak{B} \cap \mathfrak{L}_a$ for $a \in A$.

4.1.15 Definition

An *A-graded ideal* (respectively *p-ideal*) of an A-graded Lie algebra (respectively Lie p-algebra) $\mathfrak{L}$ is an ideal (respectively p-ideal) $\mathfrak{B}$ of $\mathfrak{L}$ such that $\mathfrak{B} = \sum_{a\in A} \mathfrak{B}_a$ where $\mathfrak{B}_a = \mathfrak{B} \cap \mathfrak{L}_a$ for $a \in A$.

4.1.16 Definition

An *A-homomorphism* (respectively *A-p-homomorphism*) of A-graded Lie algebras (respectively Lie p-algebras) $\mathfrak{L}$, $\mathfrak{L}'$ is a homomorphism $f: \mathfrak{L} \to \mathfrak{L}'$ of Lie algebras (respectively Lie p-algebras) such that $f(\mathfrak{L}_a) \subset \mathfrak{L}'_a$ for all $a \in A$.

4.1.17 Definition

An *A-isomorphism* (respectively *A-p-isomorphism*) is a bijective *A-homomorphism* (respectively *A-p-homomorphism*).

If $\mathfrak{L}$ is an A-graded Lie algebra (respectively Lie p-algebra) with A-graded ideal (respectively p-ideal) $\mathfrak{B}$, then $\mathfrak{L}/\mathfrak{B}$ has a unique A-graded structure such that the canonical homomorphism $\mathfrak{L} \to \mathfrak{L}/\mathfrak{B}$ is an A-homomorphism (respectively A-p-homomorphism).

4.1.18 Definition

An *A-graded Lie module* (respectively *Lie p-module*) for an A-graded Lie algebra (respectively Lie p-algebra) $\mathfrak{L}$ is a Lie module (respectively Lie p-module) $\mathfrak{V}$ for $\mathfrak{L}$ together with a decomposition $\mathfrak{V} = \sum_{a\in A} \mathfrak{V}_a$ (direct sum of subspaces) such that $\mathfrak{V}_a \mathfrak{L}_b \subset \mathfrak{V}_{ab}$ for all $a, b \in A$.

If $\mathfrak{L}$ is an A-graded Lie algebra (respectively Lie p-algebra), the Lie module (respectively Lie p-module) of $\mathfrak{L}$ afforded by ad is obviously A-graded. As we show later, it follows that the center of $\mathfrak{L}$ (the kernel of ad) is A-graded.

4.1.19 Definition

An *A-graded submodule* (respectively *p-submodule*) of an A-graded Lie module (respectively Lie p-module) $\mathfrak{M}$ of an A-graded Lie algebra (respectively Lie p-algebra) $\mathfrak{L}$ is a submodule (respectively p-submodule) $\mathfrak{N}$ of $\mathfrak{M}$ such that $\mathfrak{N} = \sum_{a \in A} \mathfrak{N}_a$ where $\mathfrak{N}_a = \mathfrak{N} \cap \mathfrak{M}_a$ for $a \in A$. (One then automatically has $\mathfrak{N}_a \mathfrak{L}_b \subset \mathfrak{N}_{ab}$ for $a, b \in A$).

4.2 *A*-Graded Lie Algebras

We continue to assume that A and B are groups, and let the identity of A be denoted 1.

4.2.1 Proposition

Let $\mathfrak{L}$ be an A-graded Lie algebra, $\mathfrak{M}$ an A-graded Lie module for $\mathfrak{L}$. Then $\mathrm{Ann}_{\mathfrak{L}} \mathfrak{M} = \{x \in \mathfrak{L} \mid \mathfrak{M}x = \{0\}\}$ is an A-graded ideal of $\mathfrak{L}$.

PROOF. Let $x \in \mathrm{Ann}_{\mathfrak{L}} \mathfrak{M}$ and $x = \sum_{a \in A} x_a$ with $x_a \in \mathfrak{L}_a$ for all $a \in A$. We show that $x_a \in \mathrm{Ann}_{\mathfrak{L}} \mathfrak{M}$ for $a \in A$. Let $m_b \in \mathfrak{M}_b$. Then $0 = m_b x = \sum m_b x_a$. Since $m_b x_a \in \mathfrak{M}_{ba}$ for all a, $m_b x_a = 0$ for all a, by the directness of the sum $\sum_{a \in A} \mathfrak{M}_{ba}$. Thus, $\mathfrak{M}x_a = \sum_{b \in A} \mathfrak{M}_b x_a = \{0\}$ and $x_a \in \mathrm{Ann}_{\mathfrak{L}} \mathfrak{M}$ for all $a \in A$.

4.2.2 Corollary

The center $\mathfrak{C}$ of an A-graded Lie algebra $\mathfrak{L}$ is A-graded.

PROOF. $\mathfrak{C} = \mathrm{Ann}_{\mathfrak{L}} \mathfrak{L}$ where $\mathfrak{L}$ is regarded as a Lie module for $\mathfrak{L}$ by way of ad.

We now describe some useful graded subalgebras of an A-graded Lie algebra $\mathfrak{L}$.

4.2.3 Definition

For B a subgroup of A, $\mathfrak{L}^B$ is the B-graded Lie algebra $\mathfrak{L}^B = \sum_{b \in B} \mathfrak{L}_b$.

4.2.4 Definition

If f is a homomorphism of groups from A to B, then $\mathfrak{L}^f$ is the B-graded Lie algebra $\mathfrak{L}^f = \sum_{b \in B} \mathfrak{L}^b$ where $\mathfrak{L}^b = \sum_{f(a) = b} \mathfrak{L}_a$ for $b \in B$. Of course, $\mathfrak{L} = \mathfrak{L}^f$ as Lie algebras.

If $B = \{1\}$ where 1 is the identity of A, then $\mathfrak{L}^B = \mathfrak{L}_1$. If B is a normal subgroup of A and $f: A \to A/B$ the canonical homomorphism, then the grading for the A/B-graded Lie algebra $\mathfrak{L}^f$ is $\mathfrak{L}^f = \mathfrak{L} = \sum_{aB \in A/B} \mathfrak{L}^{aB}$, and $\mathfrak{L}^{aB} = \sum_{ab \in aB} \mathfrak{L}_{ab}$ for $aB \in A/B$.

4.2.5 Proposition

Let $\mathfrak{L}$ be A-graded and $\mathfrak{J}, \mathfrak{J}'$ A-graded ideals of $\mathfrak{L}$. Then $\mathfrak{J}\mathfrak{J}'$ is an A-graded ideal of $\mathfrak{L}$. Consequently, the ideals $\mathfrak{J}^i$ and $\mathfrak{L}^{(i)}$ are A-graded for all i.

PROOF. $\mathfrak{J} = \sum_{a \in A} \mathfrak{J}_a$ and $\mathfrak{J}' = \sum_{a \in A} \mathfrak{J}'_a$, so that $\mathfrak{J}\mathfrak{J}' = \sum_{a, b \in A} \mathfrak{J}_a \mathfrak{J}'_b$. Since $\mathfrak{J}_a \mathfrak{J}'_b \subset \mathfrak{J}\mathfrak{J}' \cap \mathfrak{L}_{ab} = (\mathfrak{J}\mathfrak{J}')_{ab}$, $\mathfrak{J}\mathfrak{J}' = \sum_{a \in A} (\mathfrak{J}\mathfrak{J}')_a$.

4.2.6 Proposition

Let $\mathfrak{L}$ be an A-graded Lie algebra, $\mathfrak{H}$ a Cartan subalgebra of $\mathfrak{L}_1$. Then $\mathfrak{B} = \mathfrak{L}_0(\text{ad } \mathfrak{H})$ is an A-graded subalgebra of $\mathfrak{L}$ and $\mathfrak{B}_1 = \mathfrak{H}$.

PROOF. We know that $\mathfrak{B} = \mathfrak{L}_0(\text{ad } \mathfrak{H})$ is a subalgebra of $\mathfrak{L}$, by 2.4.12. Now the $\mathfrak{L}_a$ are ad $\mathfrak{H}$-stable, by 2.4.12, so that $\mathfrak{B} = (\sum \mathfrak{L}_a)_0(\text{ad } \mathfrak{H}) = \sum (\mathfrak{L}_a)_0(\text{ad } \mathfrak{H}) = \sum \mathfrak{B} \cap \mathfrak{L}_a = \sum \mathfrak{B}_a$ and $\mathfrak{B}$ is A-graded. Moreover, $\mathfrak{B}_1 = (\mathfrak{L}_1)_0(\text{ad } \mathfrak{H}) = \mathfrak{H}$ since $\mathfrak{H}$ is a Cartan subalgebra of $\mathfrak{L}_1$.

4.2.7 Theorem

Let A be a finite cyclic group, $\mathfrak{L}$ an A-graded Lie algebra. Let $\mathfrak{H}$ be a Cartan subalgebra of $\mathfrak{L}_1$. Then $\mathfrak{L}_0(\text{ad } \mathfrak{H})$ is solvable.

PROOF. Since $\mathfrak{B} = \mathfrak{L}_0(\text{ad } \mathfrak{H})$ is A-graded and $\mathfrak{B}_1 = \mathfrak{H}$, by 4.2.6, we may assume without loss of generality that $\mathfrak{L}_1 = \mathfrak{H}$ and $\mathfrak{L} = \mathfrak{L}_0(\text{ad } \mathfrak{H})$. Now what is to be proved is that $\mathfrak{L}$ is solvable. The proof is by induction on $\dim \mathfrak{L}$ and is trivial if $\dim \mathfrak{L} = 0$.

We consider two cases. Suppose first that $\mathfrak{L} \neq \mathfrak{L}^{(1)}$. Then $\mathfrak{L}^{(1)}$ is A-graded, by 4.2.5, $(\mathfrak{L}^{(1)})_1 = \mathfrak{L}^{(1)} \cap \mathfrak{H}$ is nilpotent, and $(\mathfrak{L}^{(1)})_0(\text{ad } (\mathfrak{L}^{(1)})_1) = \mathfrak{L}^{(1)}$. Since $\dim \mathfrak{L}^{(1)} < \dim \mathfrak{L}$, $\mathfrak{L}^{(1)}$ is solvable, by induction. Thus $\mathfrak{L}$ is solvable.

Now suppose that $\mathfrak{L} = \mathfrak{L}^{(1)}$. We will then show that $\mathfrak{L} = \{0\}$. For this, regard A as an additive group, let 0 denote the identity element of A, and let 1 denote a generator of A. Order A such that $a < a + 1$ for $a \neq 0$. It then follows that $a < 0$ for $a \neq 0$ and that $a + b = c$ and $a \geq c$ implies $b \geq c$. We claim that $\mathfrak{L} = \mathfrak{L}_{[a]}$ for all $a \in A$, where $\mathfrak{L}_{[a]}$ is the subalgebra of $\mathfrak{L}$ generated by $\bigcup_{b \geq a} \mathfrak{L}_b$. The proof is by induction. Clearly $\mathfrak{L} = \mathfrak{L}_{[1]}$, since $1 \leq b$ for all $b \in A$. Suppose $\mathfrak{L} = \mathfrak{L}_{[a]}$. We then show that $\mathfrak{L} = \mathfrak{L}_{[a+1]}$. Since $\mathfrak{L} = \mathfrak{L}^{(1)} = \mathfrak{L}_{[a]}\mathfrak{L}_{[a]}$, $\mathfrak{L}$ is the sum of subspaces of the form $\mathfrak{L}_{(i)} = (\ldots((\mathfrak{L}_{i_1}, \mathfrak{L}_{i_2})\mathfrak{L}_{i_3})\ldots)\mathfrak{L}_{i_r})$ where $(i) = (i_1, \ldots, i_r)$, $r \geq 2$ and $i_j \geq a$ for all j. In particular, an element of $\mathfrak{L}_a$ is a sum of elements of the form $l_{(i)} \in \mathfrak{L}_{(i)}$ where $(i) = (i_1, \ldots, i_r)$, $r \geq 2$, $i_j \geq a$ for all j and $\sum_1^r i_j = a$. For any such $l_{(i)}$, let $a_1 = \sum_1^{r-1} i_j$, $a_2 = i_r$. Then we have $a_1 + a_2 = a$ and $a_2 \geq a$, and $l_{(i)} = xy$ with $x \in \mathfrak{L}_{a_1}$ and $y \in \mathfrak{L}_{a_2}$. By a preceding observation about the order $<$, we have $a_1 \geq a$. If $a_2 = a$, then $a_1 = 0$; and if $a_1 = a$, then $a_2 = 0$. Thus either $a_1 > a$ and $a_2 > a$, in which case $l_{(i)} \in \mathfrak{L}_{[a+1]}$; or $\{a_1, a_2\} = \{0, a\}$, in which case $l_{(i)} \in \mathfrak{L}_a \text{ad } \mathfrak{L}_0$. Thus, $\mathfrak{L}_a \subset \mathfrak{L}_{[a+1]} + \mathfrak{L}_a \text{ad } \mathfrak{L}_0$. One therefore sees that, recursively, $\mathfrak{L}_a(\text{ad } \mathfrak{L}_0)^i \subset \mathfrak{L}_{[a+1]} + \mathfrak{L}_a(\text{ad } \mathfrak{L}_0)^{i+1}$ for all i, since $\mathfrak{L}_{[a+1]}$ is ad $\mathfrak{L}_0$-stable. Thus, $\mathfrak{L}_a \subset \mathfrak{L}_{[a+1]} + \mathfrak{L}_a(\text{ad } \mathfrak{L}_0)^{i+1}$ for all i. Since ad $\mathfrak{L}_0$ consists of nilpotent elements, the term $\mathfrak{L}_a(\text{ad } \mathfrak{L}_0)^{i+1}$ is $\{0\}$ for i sufficiently large, by 3.2.9. Thus, $\mathfrak{L}_a \subset \mathfrak{L}_{[a+1]}$. It follows that $\mathfrak{L} = \mathfrak{L}_{[a]} = \mathfrak{L}_{[a+1]}$. Thus, $\mathfrak{L} = \mathfrak{L}_{[b]}$ for all $b \in A$. In particular, $\mathfrak{L} = \mathfrak{L}_{[0]} = \mathfrak{L}_0$. Thus, $\mathfrak{L}$ is nilpotent by 3.2.9, since ad $\mathfrak{L}_0$ consists of nilpotent elements. Since $\mathfrak{L} = \mathfrak{L}^{(1)}$, it follows that $\mathfrak{L} = \{0\}$.

4.3 Nilpotent Lie Algebras

4.3.1 Proposition

Let $\mathfrak{N}$ be a nilpotent Lie algebra, and let $\mathfrak{V}$ be a finite-dimensional Lie module for $\mathfrak{N}$. Then $\mathfrak{V} = \mathfrak{V}_0(\mathfrak{N}) \oplus \mathfrak{V}_*(\mathfrak{N})$.

PROOF. If the ground field k is algebraically closed, this follows from 3.2.11. Now let k' be the algebraic closure of k, $\mathfrak{V}' = \mathfrak{V}_{k'}$, and $\mathfrak{N}' = \mathfrak{N}_{k'}$. Then $\mathfrak{N}'$ is nilpotent, by 2.2.3, and $\mathfrak{V}' = \mathfrak{V}'_0(\mathfrak{N}') \oplus \mathfrak{V}'_*(\mathfrak{N}')$, by 3.2.11. But $\mathfrak{V}_0(\mathfrak{N})$ and $\mathfrak{V}_*(\mathfrak{N})$ are k-forms of $\mathfrak{V}'_0(\mathfrak{N}')$ and $\mathfrak{V}'_*(\mathfrak{N}')$ respectively, by 1.5.4. Thus, $\mathfrak{V} = \mathfrak{V}_0(\mathfrak{N}) \oplus \mathfrak{V}_*(\mathfrak{N})$.

4.3.2 Proposition

Let k be infinite and $\mathfrak{N}$ a nilpotent Lie algebra. Let $\mathfrak{V}$ be a finite-dimensional module for $\mathfrak{N}$. Then there exists a $T \in \mathfrak{N}$ such that $\mathfrak{V}_0(\mathfrak{N}) = \mathfrak{V}_0(T)$ and $\mathfrak{V}_*(\mathfrak{N}) = \mathfrak{V}_*(T)$.

PROOF. Without loss of generality, we may assume that $\mathfrak{N}$ is a subalgebra of $(\mathrm{Hom}_k \mathfrak{V})_{\mathrm{Lie}}$. Take $T_1 \in \mathfrak{N}$ such that $\dim \mathfrak{V}_0(T_1)$ is minimal. Recall that $\mathfrak{V}_0(T_1)$ is $\mathfrak{N}$-stable, by 3.2.13. Suppose that $\mathfrak{V}_0(T_1) \supsetneqq \mathfrak{V}_0(\mathfrak{N})$. Then there exists $T_2 \in \mathfrak{N}$ such that $T_2|_{\mathfrak{V}_0(T_1)}$ is not nilpotent. Now $\mathfrak{V} = \mathfrak{V}_0(T_i) \oplus \mathfrak{V}_*(T_i)$ and the $\mathfrak{V}_0(T_i)$, $\mathfrak{V}_*(T_i)$ are $\mathfrak{N}$-stable ($i = 1, 2$). Let $U_i = \{aT_1 + bT_2 \mid a, b \in k, \det (aT_1 + bT_2)|_{\mathfrak{V}_*(T_i)} = 0\}$ ($i = 1, 2$). Then $T_i \in U_i$ and the U_i are open dense subsets of $kT_1 + kT_2$ ($i = 1, 2$), by A.5. Thus, $U_1 \cap U_2$ is nonempty, by A.6. Take $T \in U_1 \cap U_2$. The spaces $\mathfrak{V}_0 = \mathfrak{V}_0(T_1)$ and $\mathfrak{V}_* = \mathfrak{V}_*(T_1)$ are T-stable and $T|_{\mathfrak{V}_*}$ is nonsingular. From the decomposition $\mathfrak{V} = \mathfrak{V}_0 \oplus \mathfrak{V}_* = (\mathfrak{V}_0)_0(T) + ((\mathfrak{V}_0)_*(T) + \mathfrak{V}_*)$, we see that

$$\mathfrak{V}_0(T) = (\mathfrak{V}_0)_0(T) \subset \mathfrak{V}_0 = \mathfrak{V}_0(T_1),$$

$$\mathfrak{V}_*(T) = (\mathfrak{V}_0)_*(T) \oplus \mathfrak{V}_* \supset \mathfrak{V}_* = \mathfrak{V}_*(T_1).$$

Since $T|_{\mathfrak{V}_*(T_1)+\mathfrak{V}_*(T_2)}$ is nonsingular and $\mathfrak{V}_*(T_1) + \mathfrak{V}_*(T_2) \supsetneqq \mathfrak{V}_*(T_1)$, we have $\mathfrak{V}_*(T) \supsetneqq \mathfrak{V}_*(T_1)$. Since $\mathfrak{V} = \mathfrak{V}_0(T_1) \oplus \mathfrak{V}_*(T_1)$, it follows that $\mathfrak{V}_0(T) \subsetneqq \mathfrak{V}_0(T_1)$, a contradiction. Thus, $\mathfrak{V}_0(T_1) = \mathfrak{V}_0(\mathfrak{N})$. Since $\mathfrak{V} = \mathfrak{V}_0(T_1) \oplus \mathfrak{V}_*(T_1) = \mathfrak{V}_0(\mathfrak{N}) \oplus \mathfrak{V}_*(\mathfrak{N})$ and $\mathfrak{V}_*(T_1) \subset \mathfrak{V}_*(\mathfrak{N})$, it follows that $\mathfrak{V}_*(T_1) = \mathfrak{V}_*(\mathfrak{N})$.

We now prove an important generalization of Engel's theorem, due to N. Jacobson. We follow the notation of 4.2.

4.3.3 Definition

Let $\mathfrak{H}$ and $\mathfrak{N}$ be subsets of $\mathrm{Hom}_k \mathfrak{V}$, where $\mathfrak{V}$ is a finite-dimensional vector space over k. Then $\mathfrak{N}$ *weakly normalizes* $\mathfrak{H}$ if for each $x \in \mathfrak{N}$ and $y \in \mathfrak{H}$, $xy - \alpha yx \in \mathfrak{H}$ for some $\alpha \in k$ dependent on x and y.

4.3.4 Lemma

Let $\mathfrak{N}$, $\mathfrak{H}$, $\mathfrak{V}$ be as in 4.3.3, and suppose that $\mathfrak{N}$ weakly normalizes $\mathfrak{H}$. Then the spaces $\mathfrak{V}_0^i(\mathfrak{H})$ are $\mathfrak{N}$-stable.

PROOF. The proof is by induction on i and is trivial if $i \leq 1$. Thus, let $i > 1$ and suppose that $\mathfrak{V}_0^{i-1}(\mathfrak{H})$ is $\mathfrak{N}$-stable. Let $v \in \mathfrak{V}_0^i(\mathfrak{H})$, $x \in \mathfrak{N}$, $y \in \mathfrak{H}$. Then $(vx)y = \alpha(vy)x + vy'$ for some $y' \in \mathfrak{H}$ and some $\alpha \in k$, so that $(vx)y \in \mathfrak{V}_0^{i-1}(\mathfrak{H})\mathfrak{N} + \mathfrak{V}_0^{i-1}(\mathfrak{H}) \subset \mathfrak{V}_0^{i-1}(\mathfrak{H})$. Thus, $vx \in \mathfrak{V}_0^i(\mathfrak{H})$ for all $x \in \mathfrak{N}$, and $\mathfrak{V}_0^i(\mathfrak{H})$ is $\mathfrak{N}$-stable.

4.3.5 Definition

A subset $\mathfrak{N}$ of $\text{Hom}_k\, \mathfrak{V}$ is *weakly closed* if there exists a function $a: \mathfrak{N} \times \mathfrak{N} \to k$ such that $\mathfrak{N}$ is closed under the composition $x \circ y = xy - a(x, y)yx$.

4.3.6 Definition

The *enveloping associative algebra* $\mathfrak{N}^*$ of a subset $\mathfrak{N}$ of $\text{Hom}_k\, \mathfrak{V}$ is the intersection of all multiplicatively closed subspaces of $\text{Hom}_k\, \mathfrak{V}$ containing $\mathfrak{N}$. It is regarded as an associative algebra.

4.3.7 Theorem (Jacobson)

Let $\mathfrak{N}$ be a weakly closed subset of $\text{Hom}_k\, \mathfrak{V}$, $\mathfrak{V}$ being a finite-dimensional vector space over k. Then $\mathfrak{V}_0(\mathfrak{N}) = \mathfrak{V}_0^i(\mathfrak{N})$ for some i. If $\mathfrak{N}$ consists of nilpotent elements, then $\mathfrak{N}^*$ is nilpotent and $\mathfrak{V}$ has a basis relative to which the matrices of elements of $\mathfrak{N}^*$ are in strictly upper triangular form:

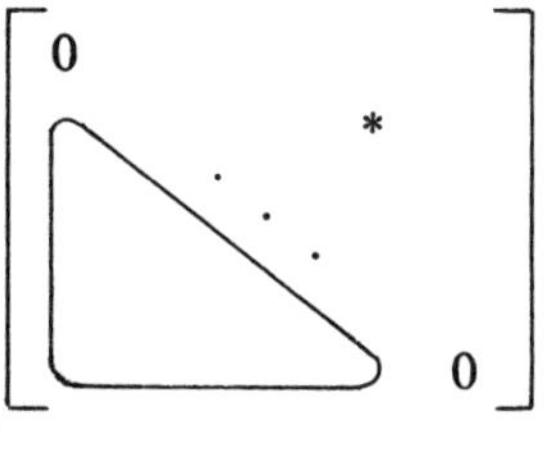

Figure 35.

PROOF. It suffices to show that $\mathfrak{V}_0(\mathfrak{N}) = \mathfrak{V}_0^i(\mathfrak{N})$ for some i. For then, if every element of $\mathfrak{N}$ is nilpotent, $\mathfrak{V} = \mathfrak{V}_0^i(\mathfrak{N})$. And if $\mathfrak{V} = \mathfrak{V}_0^i(\mathfrak{N})$, there is a basis of $\mathfrak{V}$ relative to which the matrices of elements of $\mathfrak{N}$, hence of $\mathfrak{N}^*$, are all in strictly upper triangular form. The latter implies that $\mathfrak{N}^*$ is nilpotent as an associative algebra.

In showing that $\mathfrak{V}_0(\mathfrak{N}) = \mathfrak{V}_0^i(\mathfrak{N})$ for some i, we may assume that $\mathfrak{V} = \mathfrak{V}_0(\mathfrak{N})$, for otherwise we could replace $\mathfrak{V}$ by $\mathfrak{V}_0(\mathfrak{N})$ and $\mathfrak{N}$ by $\{x|_{\mathfrak{V}_0(\mathfrak{N})} \mid x \in \mathfrak{N}\}$. Now let $a: \mathfrak{N} \times \mathfrak{N} \to k$ be a function such that $\mathfrak{N}$ is closed under the composition $x \circ y = xy - a(x, y)yx$. Let $\mathfrak{H}$ be a subset of $\mathfrak{N}$ such that $\mathfrak{H} \circ \mathfrak{H} \subset \mathfrak{H}$ and $\mathfrak{V} = \mathfrak{V}_0^i(\mathfrak{H})$ for some i, and choose $\mathfrak{H}$ with $\dim \mathfrak{H}^*$ maximal. We claim that $\mathfrak{N} \subset \mathfrak{H}^*$, whence $\mathfrak{V}_0^i(\mathfrak{N}) \supset \mathfrak{V}_0^i(\mathfrak{H}^*) = \mathfrak{V}_0^i(\mathfrak{H}) = \mathfrak{V}$ and $\mathfrak{V}_0^i(\mathfrak{N}) = \mathfrak{V}$ for some i. Suppose not and let $x \in \mathfrak{N} - \mathfrak{H}^*$. Let $x(y_1, \ldots, y_n) = (\ldots((x \circ y_1) \circ y_2) \ldots) \circ y_n$, which is a linear combination of terms of the form $y_{i_1} \ldots y_{i_m} x y_{i_{m+1}} \ldots y_{i_n}$. Since $\mathfrak{H}^*$ is nilpotent, there exists n such that $x(y_1, \ldots, y_n) = 0$ for all $y_1, \ldots, y_n \in \mathfrak{H}$. Taking $n \geq 1$ minimal such that $x(y_1, \ldots, y_n) \in \mathfrak{H}^*$ for all $y_1, \ldots, y_n \in \mathfrak{H}$, there exist $y_1, \ldots, y_{n-1} \in \mathfrak{H}$ such that $z = x(y_1, \ldots, y_{n-1}) \in \mathfrak{N} - \mathfrak{H}^*$. Now $z \circ y \in \mathfrak{H}^* \cap \mathfrak{N} = \mathfrak{H}$ for $y \in \mathfrak{H}$, so that $\mathfrak{H}$ is weakly normalized by z. Thus, the chain $\{0\} = \mathfrak{V}_0^0(\mathfrak{H}) \subset \ldots \subset \mathfrak{V}_0^i(\mathfrak{H}) = \mathfrak{V}$ is z-stable and, since z is nilpotent, has a refinement $\{0\} = \mathfrak{V}^0 \subset \ldots \subset \mathfrak{V}^j = \mathfrak{V}$ such that $\mathfrak{V}^m z \subset \mathfrak{V}^{m-1}$ for $1 \leq m \leq j$. One automatically has $\mathfrak{V}^m \mathfrak{H} \subset \mathfrak{V}^{m-1}$ for $1 \leq m \leq j$, since $\mathfrak{V}_0^m(\mathfrak{H})\mathfrak{H} \subset \mathfrak{V}_0^{m-1}(\mathfrak{H})$ for $1 \leq m \leq i$. It follows that $\mathfrak{V}_0^j(\mathfrak{H}') = \mathfrak{V}$ where $\mathfrak{H}' = \mathfrak{H} \cup \{z\}$. But $\mathfrak{H}' \circ \mathfrak{H}' \subset \mathfrak{H}'$ and $\dim \mathfrak{H}'^* > \dim \mathfrak{H}^*$, a contradiction. Thus, $\mathfrak{N} \subset \mathfrak{H}^*$ as asserted.

We now give some nilpotency and solvability criteria for certain A-graded Lie algebras.

4.3.8 Theorem

Let A be a group, $\mathfrak{L}$ an A-graded Lie algebra, $\mathfrak{V}$ an A-graded Lie module for $\mathfrak{L}$. Let B be a torsion-free subgroup of A. Then $\mathfrak{V} = \mathfrak{V}_0(\mathfrak{L}_1)$ iff $\mathfrak{V} = \mathfrak{V}_0(\mathfrak{L}^B)$.

PROOF. Let f be the representation of $\mathfrak{L}$ afforded by $\mathfrak{V}$. Assume that $\mathfrak{V} = \mathfrak{V}_0(\mathfrak{L}_1)$. We must show that $f(x)$ is nilpotent for $x \in \mathfrak{L}^B$. The assertion then follows from 3.2.9. We assume first that $x \in \mathfrak{L}_b$ with $b \in B$, and show that $f(x)$ is nilpotent. Note that $\mathfrak{V}_a(f(x))^n \subset \mathfrak{V}_{ab^n}$ for $a \in A$ and n any positive integer. Since B is torsion free, the $ab^i (i = 1, 2, \ldots)$ are distinct and $\mathfrak{V}_{ab^n} = \{0\}$ for some n. Thus, $\mathfrak{V}_a(f(x))^n = 0$. It follows that $f(x)$ is nilpotent, since $\mathfrak{V} = \sum_{a \in A} \mathfrak{V}_a$. Thus, $\bigcup_{b \in B} f(\mathfrak{L}_b)$ consists of nilpotent transformations of $\mathfrak{V}$. Since it is closed under [,], 4.3.7 applies and $f(\mathfrak{L}^B)$ consists of nilpotent transformations.

4.3.9 Theorem

Let A be a group, $\mathfrak{L}$ an A-graded Lie p-algebra, $\mathfrak{V}$ an A-graded Lie p-module for $\mathfrak{L}$. Let B be a (possibly infinite) p-subgroup of A. Then $\mathfrak{V} = \mathfrak{V}_0(\mathfrak{L}_1)$ iff $\mathfrak{V} = \mathfrak{V}_0(\mathfrak{L}^B)$.

PROOF. Let $\mathfrak{V} = \mathfrak{V}_0(\mathfrak{L}_1)$. We show that $\mathfrak{V} = \mathfrak{V}_0(\mathfrak{L}^B)$ in the same way as in the proof of 4.3.8, except that we argue as follows in showing that $f(x)$ is nilpotent for $x \in \mathfrak{L}_b$, $b \in B$. Let p^e be the order of b and let $a \in A$. Then

$$\mathfrak{V}_a f(x)^{p^e} \subset \mathfrak{V}_{ab^{p^e}} = \mathfrak{V}_a.$$

Letting $y = x^{p^e}$, $\mathfrak{V}_a f(y) = \mathfrak{V}_a f(x)^{p^e} \subset \mathfrak{V}_a$ and $\mathfrak{V}_a$ is $f(y)$-stable for $a \in A$. Decompose y into $y = \sum_{c \in A} y_c$ where $y_c \in \mathfrak{L}_c$ for $c \in A$. Let $v_a \in \mathfrak{V}_a$. Then $\sum_{c \in A} v_a y_c = v_a y \in \mathfrak{V}_a$, so $v_a y_c = 0$ for $c \neq 1$ by the directness of $\mathfrak{V} = \sum \mathfrak{V}_{ac}$. Thus, $v_a y = v_a y_1$ for all a and $v_a \in \mathfrak{V}_a$ and, consequently, $vy = vy_1$ for all $v \in \mathfrak{V}$. Thus, $f(y) = f(y_1) \in f(\mathfrak{L}_1)$ so that $f(y)$ is nilpotent. But then $f(x)$ is nilpotent, since $f(y) = f(x)^{p^e}$.

4.3.10 Corollary

Let A be a group, $\mathfrak{L}$ an A-graded Lie p-algebra, $\mathfrak{V}$ an A-graded Lie p-module for $\mathfrak{L}$. Suppose that $A = \bigcup_{i=1}^{\infty} A_i$ where $\{1\} = A_1 < A_2 < \ldots$ is a (finite or infinite) sequence of subgroups of A such that, for each i, A_i is normal in A_{i+1} and A_{i+1}/A_i either torsion free or a p-group. Then $\mathfrak{V} = \mathfrak{V}_0(\mathfrak{L}_1)$ iff $\mathfrak{V} = \mathfrak{V}_0(\mathfrak{L})$.

PROOF. Assume that $\mathfrak{V} = \mathfrak{V}_0(\mathfrak{L}_1)$. It suffices to show that $\mathfrak{V} = \mathfrak{V}_0(\mathfrak{L}^{A_i})$ implies that $\mathfrak{V} = \mathfrak{V}_0(\mathfrak{L}^{A_{i+1}})$ for all i. For then $\mathfrak{V} = \mathfrak{V}_0(\mathfrak{L}^{A_i})$ for all i, by induction, and $\mathfrak{V} = \mathfrak{V}_0(\mathfrak{L})$. Thus, let $\mathfrak{V} = \mathfrak{V}_0(\mathfrak{L}^{A_i})$. Let $f: A_{i+1} \to A_{i+1}/A_i$ be the canonical homomorphism and let $f(A_{i+1}) = B$, $f(1) = 1_B$. Then $\mathfrak{L}$ and $\mathfrak{V}$ are B-graded by $\mathfrak{L}^b = \sum_{a \in f^{-1}(b)} \mathfrak{L}_a$ and $\mathfrak{V}^b = \sum_{a \in f^{-1}(b)} \mathfrak{V}_a$. Note that $\mathfrak{L}^{1_B} = \mathfrak{L}^{A_i}$ and $\mathfrak{L}^B = \mathfrak{L}^{A_{i+1}}$. Now $\mathfrak{V} = \mathfrak{V}_0(\mathfrak{L}^{A_i}) = \mathfrak{V}_0(\mathfrak{L}^{1_B})$ and, since B is a p-group or torsion free, $\mathfrak{V} = \mathfrak{V}_0(\mathfrak{L}^B) = \mathfrak{V}_0(\mathfrak{L}^{A_{i+1}})$, by 4.3.8 and 4.3.9.

4.3.11 Theorem

Let A be a torsion-free group (respectively p-group), $\mathfrak{L}$ an A-graded Lie algebra (respectively Lie algebra of characteristic p such that ad $\mathfrak{L}$ is closed under pth powers). Let $\mathfrak{H}$ be a Cartan subalgebra of $\mathfrak{L}_1$. Then $\mathfrak{L}_0(\text{ad}\,\mathfrak{H})$ is a Cartan subalgebra of $\mathfrak{L}$. In particular, $\mathfrak{H}$ is contained in a unique Cartan subalgebra $\hat{\mathfrak{H}}$ of $\mathfrak{L}$.

PROOF. Let $\mathfrak{H}$ be a Cartan subalgebra of $\mathfrak{L}_1$, $\mathfrak{B} = \mathfrak{L}_0(\text{ad}\,\mathfrak{H})$. Then $\mathfrak{B}$ is an A-graded subalgebra and $\mathfrak{B}_1 = \mathfrak{H}$, by 4.2.6. If $\mathfrak{L}$ is of characteristic p and ad $\mathfrak{L}$ closed under pth powers, then $\text{ad}_{\mathfrak{B}}\mathfrak{B}$ is closed under pth powers, as we verify in the following paragraph. Furthermore, ad $\mathfrak{B}$ is A-graded, by 4.2, and $\mathfrak{B} = \sum_{a\in A}\mathfrak{B}_a$ is an A-graded Lie module (respectively Lie p-module) for $\text{ad}_{\mathfrak{B}}\,\mathfrak{B}$, by 4.2. Since $\mathfrak{H} = \mathfrak{B}_1$ and $\mathfrak{B} = \mathfrak{L}_0(\text{ad}\,\mathfrak{H})$, $\mathfrak{B} = \mathfrak{B}_0(\text{ad}\,\mathfrak{B}_1)$. Thus, $\mathfrak{B} = \mathfrak{B}_0(\text{ad}\,\mathfrak{B})$, by 4.3.8 and 4.3.9, and $\mathfrak{B}$ is nilpotent by 3.2.6. Since $\mathfrak{H} \subset \mathfrak{B}$, $\mathfrak{B} \subset \mathfrak{L}_0(\text{ad}\,\mathfrak{B}) \subset \mathfrak{L}_0(\text{ad}\,\mathfrak{H}) = \mathfrak{B}$ and $\mathfrak{B} = \mathfrak{L}_0(\text{ad}\,\mathfrak{B})$. Thus, $\mathfrak{B}$ is a Cartan subalgebra of $\mathfrak{L}$.

We now return to showing that $\text{ad}_{\mathfrak{B}}\,\mathfrak{B}$ is closed under pth powers, assuming that $\mathfrak{L}$ is of characteristic p and ad $\mathfrak{L}$ closed under pth powers. Note first that $\mathfrak{N}_{\mathfrak{L}}(\mathfrak{B})\mathfrak{H} \subset \mathfrak{B} = \mathfrak{L}_0(\text{ad}\,\mathfrak{H})$ since $\mathfrak{H} \subset \mathfrak{B}$, so that $\mathfrak{N}_{\mathfrak{L}}(\mathfrak{B}) \subset \mathfrak{L}_0(\text{ad}\,\mathfrak{H}) = \mathfrak{B}$. If $x \in \mathfrak{B}$ and $(\text{ad}\,x)^p = \text{ad}\,y \in \text{ad}\,\mathfrak{L}$, then $\mathfrak{B}\,\text{ad}\,y \subset \mathfrak{B}$ so that $y \in \mathfrak{N}_{\mathfrak{L}}(\mathfrak{B}) \subset \mathfrak{B}$ and $\text{ad}_{\mathfrak{B}}\,y \in \text{ad}_{\mathfrak{B}}\,\mathfrak{B}$. Thus, $(\text{ad}_{\mathfrak{B}}\,x)^p \in \text{ad}_{\mathfrak{B}}\,\mathfrak{B}$ for $x \in \mathfrak{B}$.

4.3.12 Corollary (Jacobson)

Let $\mathfrak{L}$ be a Lie algebra (respectively Lie algebra of characteristic p such that ad $\mathfrak{L}$ is closed under pth powers). Let $\mathfrak{L}$ have a nonsingular derivation D. Then $\mathfrak{L}$ is nilpotent.

PROOF. The hypothesis is preserved under ascent, the conclusion under descent. Thus, we may assume that k is algebraically closed. Now $\mathfrak{L} = \sum_{\alpha\in k}\mathfrak{L}_\alpha(D)$ and $\mathfrak{L}$ is k-graded, by 4.1.12. Since $\mathfrak{L}_0(D) = \{0\}$ and since $\mathfrak{L}_0(D)$ is the $\mathfrak{L}_1$ of 4.3.8 and 4.3.9, $\mathfrak{L}$ is nilpotent.

4.3.13 Theorem

Let A be a group having a normal torsion-free subgroup C such that A/C

is cyclic. Let $\mathfrak{L}$ be an A-graded Lie algebra, $\mathfrak{H}$ a Cartan subalgebra of $\mathfrak{L}_1$. Then $\mathfrak{L}_0(\operatorname{ad}\mathfrak{H})$ is solvable.

PROOF. Let $\mathfrak{B} = \mathfrak{L}_0(\operatorname{ad}\mathfrak{H})$. Then $\mathfrak{B}$ is an A-graded subalgebra and $\mathfrak{B}_1 = \mathfrak{H}$, by 4.2.6. Now $\mathfrak{B} = \mathfrak{B}_0(\operatorname{ad}\mathfrak{B}_1)$ since $\mathfrak{B} = \mathfrak{L}_0(\operatorname{ad}\mathfrak{H})$. Thus, $\mathfrak{B} = \mathfrak{B}_0(\operatorname{ad}\mathfrak{B}^C)$, by 4.3.8. But $\mathfrak{B}$ is A/C-graded, by 4.2.4, and $\mathfrak{B}^C = \mathfrak{B}^e$ where e is the identity of A/C. Thus, $\mathfrak{B} = \mathfrak{B}_0(\operatorname{ad}\mathfrak{B}^e)$. Since A/C is cyclic, it follows that $\mathfrak{B}$ is solvable, by 4.2.7.

We use this theorem in 4.4.6 to describe properties of Cartan subalgebras of the fixed-point subalgebra of a semisimple automorphism of $\mathfrak{L}$.

4.4 Cartan Subalgebras

Cartan subalgebras play a crucial role in the theory of Lie algebras of characteristic 0, and they appear also to be one of the most important tools for studying Lie algebras of characteristic p. Throughout the section, $\mathfrak{L}$ is a finite-dimensional Lie algebra over k and Lie modules for $\mathfrak{L}$ are finite dimensional.

We begin with a description of the Cartan subalgebras of a solvable Lie algebra, since this case is particularly clear-cut.

4.4.1 Cartan Subalgebras of a Solvable Lie Algebra

4.4.1.1 Theorem

Let $\mathfrak{L}$ be a Lie algebra such that the ideal $\mathfrak{L}^\infty = \bigcap_{i=1}^\infty \mathfrak{L}^i$ is Abelian. Then

1. $\mathfrak{L}$ has a Cartan subalgebra;
2. $\mathfrak{H}$ is a Cartan subalgebra of $\mathfrak{L}$ iff $\mathfrak{H}$ is a subalgebra of $\mathfrak{L}$ such that $\mathfrak{L} = \mathfrak{H} \oplus \mathfrak{L}^\infty$;
3. The group $\exp \operatorname{ad}\mathfrak{L}^\infty = \{1 + \operatorname{ad} a \mid a \in \mathfrak{L}^\infty\}$ of automorphisms of $\mathfrak{L}$ acts simply transitively on the set of Cartan subalgebras of $\mathfrak{L}$.

PROOF. Some of the motivation for the following proof comes from ideas used in 3.8.4.

We first prove 2. Suppose that $\mathfrak{H}$ is a Cartan subalgebra of $\mathfrak{L}$. Then $\mathfrak{L} = \mathfrak{H} \oplus \mathfrak{L}_*$ where $\mathfrak{L}_* = \mathfrak{L}_*(\operatorname{ad} \mathfrak{H})$. Now $\mathfrak{L}_*\mathfrak{H} = \mathfrak{L}_*$, so $\mathfrak{L}_* \subset \bigcap \mathfrak{L}^i = \mathfrak{L}^\infty$. Thus, $\mathfrak{L}_*$ is an Abelian ideal of $\mathfrak{L}$, since $\mathfrak{L}_*\mathfrak{H} \subset \mathfrak{L}_*$, $\mathfrak{L}_*\mathfrak{L}_* = \{0\}$ and $\mathfrak{L} = \mathfrak{H} \oplus \mathfrak{L}_*$. But $\mathfrak{L}/\mathfrak{L}_* \simeq \mathfrak{H}$ is nilpotent. Thus, $\mathfrak{L}^\infty \subset \mathfrak{L}_*$. Therefore, $\mathfrak{L}^\infty = \mathfrak{L}_*$ and $\mathfrak{L} = \mathfrak{H} \oplus \mathfrak{L}^\infty$. Conversely, let $\mathfrak{H}$ be a subalgebra of $\mathfrak{L}$ such that $\mathfrak{L} = \mathfrak{H} \oplus \mathfrak{L}^\infty$. Then $\mathfrak{H}$ is nilpotent, since it is isomorphic to $\mathfrak{L}/\mathfrak{L}^\infty$. Let $\mathfrak{B} = (\mathfrak{L}^\infty)_*(\operatorname{ad} \mathfrak{H})$. Since $\mathfrak{L} = \mathfrak{H} \oplus \mathfrak{L}^\infty$ and $\mathfrak{L}^\infty$ is Abelian, $\mathfrak{B}$ is an ideal of $\mathfrak{L}$. Now for $x = h + a$ with $h \in \mathfrak{H}$ and $a \in \mathfrak{L}^\infty$, $\mathfrak{L}_*(\operatorname{ad} x) = (\mathfrak{L}^\infty)_*(\operatorname{ad} x) = (\mathfrak{L}^\infty)_*(\operatorname{ad} h) \subset \mathfrak{B}$. Thus, $\mathfrak{L}/\mathfrak{B}$ is nilpotent, by 3.2.6, and $\mathfrak{L}^\infty \subset \mathfrak{B}$. Thus, $\mathfrak{B} = \mathfrak{L}^\infty$ and $\mathfrak{H} = \mathfrak{L}_0(\operatorname{ad} \mathfrak{H})$. Thus, $\mathfrak{H}$ is a Cartan subalgebra of $\mathfrak{L}$.

We next prove 1 and 3 by induction on $\dim \mathfrak{L}$. If $\dim \mathfrak{L} = 1$ or $\mathfrak{L}$ is nilpotent, these are trivial. Now let $\mathfrak{L}$ be nonnilpotent. Choose $x \in \mathfrak{L}$ such that $\operatorname{ad} x$ is not nilpotent. Then the proper subalgebra $\mathfrak{B} = \mathfrak{L}_0(\operatorname{ad} x)$ has a Cartan subalgebra $\mathfrak{H}$ by induction, since $\mathfrak{B}^\infty \subset \mathfrak{L}^\infty$ and $\mathfrak{B}^\infty$ is Abelian. Now $\mathfrak{B} = \mathfrak{H} \oplus \mathfrak{B}^\infty$ by 2, and $\mathfrak{L} = \mathfrak{B} \oplus \mathfrak{L}_*(\operatorname{ad} x) = (\mathfrak{H} \oplus \mathfrak{B}^\infty) \oplus \mathfrak{L}_*(\operatorname{ad} x) = \mathfrak{H} \oplus \mathfrak{L}^\infty$, since $\mathfrak{L}_*(\operatorname{ad} x) \subset \mathfrak{L}^\infty$ and $\mathfrak{L}/\mathfrak{J}$ is nilpotent, where $\mathfrak{J} = \mathfrak{B}^\infty \oplus \mathfrak{L}_*(\operatorname{ad} x)$. Thus, $\mathfrak{H}$ is a Cartan subalgebra of $\mathfrak{L}$, by 2. This proves 1. For 3, let $\mathfrak{L} = \mathfrak{H} \oplus \mathfrak{L}^\infty = \mathfrak{H}' \oplus \mathfrak{L}^\infty$ where $\mathfrak{H}$, $\mathfrak{H}'$ are Cartan subalgebras of $\mathfrak{L}$. We can take $y \in \mathfrak{H}'$ such that $\operatorname{ad} y$ is not nilpotent, since $\mathfrak{L}_0(\operatorname{ad} \mathfrak{H}') = \mathfrak{H}' \neq \mathfrak{L}$. Now $y = x + a \in \mathfrak{H} \oplus \mathfrak{L}^\infty$ with $x \in \mathfrak{H}$ and $a \in \mathfrak{L}^\infty$. Let $a = a_0 + bx \in (\mathfrak{L}^\infty)_0(\operatorname{ad} x) \oplus (\mathfrak{L}^\infty)_*(\operatorname{ad} x)$ where $a_0 \in (\mathfrak{L}^\infty)_0(\operatorname{ad} x)$ and $b \in (\mathfrak{L}^\infty)_*(\operatorname{ad} x)$. Letting $g = \exp(\operatorname{ad}(-b)) = 1 - \operatorname{ad} b$, then $y = x + a_0 + bx = (x + a_0)g$. Letting $\mathfrak{B} = \mathfrak{L}_0(\operatorname{ad} x)$, we see that $\mathfrak{B}$ contains $x + a_0$, and $\mathfrak{B} \subset \mathfrak{L}_0(\operatorname{ad}(x + a_0))$ since $\mathfrak{B}_*(\operatorname{ad}(x + a_0)) = (\mathfrak{B}^\infty)_*(\operatorname{ad}(x + a_0)) = (\mathfrak{B}^\infty)_*(\operatorname{ad} x) = \{0\}$. Thus, $\mathfrak{B}g$ contains $(x + a_0)g = y$, and $\mathfrak{B}g \subset \mathfrak{L}_0(\operatorname{ad} y)$. But then $\mathfrak{H}'$ and $\mathfrak{H}g$ are both Cartan subalgebras contained in the proper subalgebra $\mathfrak{L}_0(\operatorname{ad} y)$, so $\mathfrak{H}' = \mathfrak{H}gh$ for some $h \in \exp \mathfrak{L}_0(\operatorname{ad} y)^\infty$, by the induction hypothesis. Thus, the group $\exp \operatorname{ad} \mathfrak{L}^\infty$ acts transitively on the set of Cartan subalgebras of $\mathfrak{L}$. It remains to show that $\mathfrak{H} \exp \operatorname{ad} a = \mathfrak{H}$ with $a \in \mathfrak{L}^\infty$ iff $a = 0$. But this is true because $\mathfrak{H} \exp \operatorname{ad} a = \mathfrak{H}$ implies that $a \in \mathfrak{N}_{\mathfrak{L}}(\mathfrak{H}) = \mathfrak{H}$, so $a \in \mathfrak{H} \cap \mathfrak{L}^\infty = \{0\}$.

4.4.1.2 Corollary

Let $\mathfrak{L}$ be solvable. Then $\mathfrak{L}$ has a Cartan subalgebra.

PROOF. The proof is by induction on dim $\mathfrak{L}$ and is trivial if dim $\mathfrak{L} = 1$ or $\mathfrak{L}$ is nilpotent. Thus, let $\mathfrak{L}$ be nonnilpotent. Let $\mathfrak{A} = \mathfrak{L}^{(i)}$ where $\mathfrak{L}^{(i)} \neq \{0\}$ and $\mathfrak{L}^{(i+1)} = \{0\}$. Then $\mathfrak{A}$ is a nonzero Abelian ideal and $\mathfrak{L}/\mathfrak{A}$ has a Cartan subalgebra $\mathfrak{B}/\mathfrak{A}$ ($\mathfrak{B} \supset \mathfrak{A}$), by induction. Let $\mathfrak{H}$ be a Cartan subalgebra of $\mathfrak{B}$. We show that $\mathfrak{H}$ is a Cartan subalgebra of $\mathfrak{L}$. This is clear from 4.4.5.1, or by the following argument. Letting $f: \mathfrak{L} \to \mathfrak{L}/\mathfrak{A}$ be the canonical homomorphism, we have $f(\mathfrak{H}) = f(\mathfrak{B}_0(\text{ad }\mathfrak{H})) = f(\mathfrak{B})_0(\text{ad } f(\mathfrak{H}))$, as in 3.8.4 or 4.4.4.5, and $f(\mathfrak{H})$ is a Cartan subalgebra of $f(\mathfrak{B}) = \mathfrak{B}/\mathfrak{A}$. Thus, $f(\mathfrak{H}) = (\mathfrak{B}/\mathfrak{A})_0(f(\mathfrak{H})) = \mathfrak{B}/\mathfrak{A}$, since $\mathfrak{B}/\mathfrak{A}$ is nilpotent. Now $f(\mathfrak{L}_0(\text{ad }\mathfrak{H})) = f(\mathfrak{L})_0(\text{ad } f(\mathfrak{H})) = (\mathfrak{L}/\mathfrak{A})_0(\text{ad }\mathfrak{B}/\mathfrak{A}) = \mathfrak{B}/\mathfrak{A}$, and $\mathfrak{L}_0(\text{ad }\mathfrak{H}) \subset \mathfrak{B}$. It follows that $\mathfrak{L}_0(\text{ad }\mathfrak{H}) = \mathfrak{B}_0(\text{ad }\mathfrak{H}) = \mathfrak{H}$, and $\mathfrak{H}$ is a Cartan subalgebra of $\mathfrak{L}$.

4.4.2 Quasi-Regular Elements and Subalgebras

4.4.2.1 Definition
A subalgebra $\mathfrak{N}$ of Der $\mathfrak{L}$ is *quasi-regular* in $\mathfrak{L}$ if $\mathfrak{N}$ is nilpotent and $\mathfrak{L}_0(\text{ad }\mathfrak{N})$ is nilpotent.

4.4.2.2 Definition
A subalgebra $\mathfrak{N}$ of $\mathfrak{L}$ is *quasi-regular* in $\mathfrak{L}$ if ad $\mathfrak{N}$ is quasi-regular in $\mathfrak{L}$.

4.4.2.3 Definition
An element x of $\mathfrak{L}$ is *quasi-regular* in $\mathfrak{L}$ if $\mathfrak{L}_0(\text{ad } x)$ is nilpotent (that is, if the subalgebra kx is quasi-regular in $\mathfrak{L}$).

4.4.2.4 Proposition
Let k be infinite. A nilpotent subalgebra $\mathfrak{N}$ of $\mathfrak{L}$ is quasi-regular in $\mathfrak{L}$ iff $\mathfrak{N}$ contains a quasi-regular element.

PROOF. One direction is trivial. Suppose, conversely, that $\mathfrak{N}$ is quasi-regular. Then there exists ad $x \in$ ad $\mathfrak{N}$ such that $\mathfrak{L}_0(\text{ad }\mathfrak{N}) = \mathfrak{L}_0(\text{ad } x)$, by 4.3.2. Such an x is quasi-regular in $\mathfrak{L}$.

4.4.2.5 Proposition
Let $\mathfrak{N}$ be a quasi-regular subalgebra of Der $\mathfrak{L}$. Let $\mathfrak{H} = \mathfrak{L}_0(\mathfrak{N})$. Then $\mathfrak{H}$ is nilpotent and $\mathfrak{N}_{\mathfrak{L}}(\mathfrak{H}) = \mathfrak{H} + \mathfrak{C}_{\mathfrak{L}}(\mathfrak{H})$.

PROOF. $\mathfrak{H}$ is nilpotent, by the definition of quasi-regularity. Now $\mathfrak{L} = \mathfrak{L}_0(\mathfrak{N}) \oplus \mathfrak{L}_*(\mathfrak{N}) = \mathfrak{H} \oplus \mathfrak{L}_*(\mathfrak{N})$ and $\mathfrak{H}\mathfrak{L}_*(\mathfrak{N}) \subset \mathfrak{L}_*(\mathfrak{N})$. Let $x \in \mathfrak{N}_{\mathfrak{L}}(\mathfrak{H})$. Then $x = h + y$ where $h \in \mathfrak{H}$ and $y \in \mathfrak{L}_*(\mathfrak{N})$. Now $y = x - h \in \mathfrak{N}_{\mathfrak{L}}(\mathfrak{H})$, so $\mathfrak{H}y \subset \mathfrak{L}_*(\mathfrak{N}) \cap \mathfrak{H} = \{0\}$ and $y \in \mathfrak{C}_{\mathfrak{L}}(\mathfrak{H})$. Thus, $x \in \mathfrak{H} + \mathfrak{C}({}_{\mathfrak{L}}\mathfrak{H})$ and $\mathfrak{N}_{\mathfrak{L}}(\mathfrak{H}) = \mathfrak{H} + \mathfrak{C}_{\mathfrak{L}}(\mathfrak{H})$.

4.4.2.6 Proposition

A nilpotent subalgebra $\mathfrak{H}$ of $\mathfrak{L}$ is quasi-regular iff $\mathfrak{L}_0(\text{ad}\ \mathfrak{H})$ is a Cartan subalgebra of $\mathfrak{L}$.

PROOF. One direction is trivial. Conversely, let $\mathfrak{H}$ be a quasi-regular subalgebra of $\mathfrak{L}$ and $\mathfrak{B} = \mathfrak{L}_0(\text{ad}\ \mathfrak{H})$. Then $\mathfrak{H} \subset \mathfrak{B}$, so that $\mathfrak{L}_0(\text{ad}\ \mathfrak{B}) \subset \mathfrak{L}_0(\text{ad}\ \mathfrak{H}) = \mathfrak{B}$. But $\mathfrak{B} \subset \mathfrak{L}_0(\text{ad}\ \mathfrak{B})$, since $\mathfrak{B}$ is nilpotent. Thus, $\mathfrak{B} = \mathfrak{L}_0(\text{ad}\ \mathfrak{B})$ and $\mathfrak{B}$ is a Cartan subalgebra of $\mathfrak{L}$.

4.4.2.7 Corollary

An element x of $\mathfrak{L}$ is quasi-regular in $\mathfrak{L}$ iff $\mathfrak{L}_0(\text{ad}\ x)$ is a Cartan subalgebra of $\mathfrak{L}$.

PROOF. x is quasi-regular iff the algebra kx is quasi-regular.

4.4.2.8 Corollary

A quasi-regular element x (respectively quasi-regular subalgebra $\mathfrak{H}$) of $\mathfrak{L}$ is contained in a unique Cartan subalgebra $\hat{x}$ (respectively $\hat{\mathfrak{H}}$) of $\mathfrak{L}$.

PROOF. Take $\hat{x} = \mathfrak{L}_0(\text{ad}\ x)$ (respectively $\hat{\mathfrak{H}} = \mathfrak{L}_0(\text{ad}\ \mathfrak{H})$), a Cartan subalgebra of $\mathfrak{L}$, by 4.4.2.6. If $\mathfrak{N}$ were a Cartan subalgebra of $\mathfrak{L}$ containing x (respectively $\mathfrak{H}$), then $\mathfrak{N} \subset \mathfrak{L}_0(\text{ad}\ x) = \hat{x}$ (respectively $\mathfrak{N} \subset \mathfrak{L}_0(\text{ad}\ \mathfrak{H}) = \hat{\mathfrak{H}}$). But then $\mathfrak{N} = \hat{x}$ (respectively $\mathfrak{N} = \hat{\mathfrak{H}}$), since $\hat{x} \subset \mathfrak{L}_0(\text{ad}\ \mathfrak{N}) = \mathfrak{N}$ (respectively $\hat{\mathfrak{H}} \subset \mathfrak{L}_0(\text{ad}\ \mathfrak{N}) = \mathfrak{N}$) by the nilpotency of $\hat{x}$ (respectively $\hat{\mathfrak{H}}$).

4.4.2.9 Theorem

Let $\mathfrak{H}$ be a nilpotent subalgebra of $\mathfrak{L}$. Then the following conditions are equivalent:

1. $\mathfrak{H}$ is a Cartan subalgebra of $\mathfrak{L}$;

2. $\mathfrak{H}$ is a maximal quasi-regular subalgebra of $\mathfrak{L}$;
3. $\mathfrak{H} = \mathfrak{N}_{\mathfrak{L}}(\mathfrak{H})$.

PROOF. We show that $1 \Rightarrow 2 \Rightarrow 3 \Rightarrow 1$. Suppose first that $\mathfrak{H}$ is a Cartan subalgebra of $\mathfrak{L}$, that is, $\mathfrak{H} = \mathfrak{L}_0(\text{ad } \mathfrak{H})$. Then $\mathfrak{H}$ is quasi-regular. If $\mathfrak{H}'$ is quasi-regular and $\mathfrak{H}' \supset \mathfrak{H}$, then $\mathfrak{H}' \subset \mathfrak{L}_0(\text{ad } \mathfrak{H}') \subset \mathfrak{L}_0(\text{ad } \mathfrak{H}) = \mathfrak{H}$ and $\mathfrak{H}' = \mathfrak{H}$. Thus, $\mathfrak{H}$ is a maximal quasi-regular subalgebra of $\mathfrak{L}$ and $1 \Rightarrow 2$. Next, suppose that $\mathfrak{H}$ is maximal quasi-regular. Then $\mathfrak{L}_0(\text{ad } \mathfrak{H}) = \mathfrak{H}'$ is nilpotent and $\mathfrak{H}' \supset \mathfrak{H}$. Thus, $\mathfrak{L}_0(\text{ad } \mathfrak{H}') \subset \mathfrak{L}_0(\text{ad } \mathfrak{H}) = \mathfrak{H}'$ and $\mathfrak{L}_0(\text{ad } \mathfrak{H}')$ is nilpotent. Thus, $\mathfrak{H}'$ is quasi-regular. Since $\mathfrak{H}$ is maximal quasi-regular, $\mathfrak{H} = \mathfrak{H}' = \mathfrak{L}_0(\text{ad } \mathfrak{H})$. Thus, $\mathfrak{H} = \mathfrak{N}_{\mathfrak{L}}(\mathfrak{H})$ and $2 \Rightarrow 3$. Finally, suppose that $\mathfrak{H} = \mathfrak{N}_{\mathfrak{L}}(\mathfrak{H})$. By 3.2.9, $\mathfrak{L}_0(\text{ad } \mathfrak{H}) = \mathfrak{L}_0^i(\text{ad } \mathfrak{H})$ for some i. Suppose that $\mathfrak{L}_0(\text{ad } \mathfrak{H}) \supsetneqq \mathfrak{H}$. Choosing m maximal such that $\mathfrak{H} \supset \mathfrak{L}_0^m(\text{ad } \mathfrak{H})$, we can find $x \in \mathfrak{L}_0^{m+1}(\text{ad } \mathfrak{H}) - \mathfrak{H}$. Now $x\mathfrak{H} \subset \mathfrak{L}_0^m(\text{ad } \mathfrak{H}) \subset \mathfrak{H}$ and $\mathfrak{N}_{\mathfrak{L}}(\mathfrak{H}) \neq \mathfrak{H}$, a contradiction. Thus $\mathfrak{L}_0(\text{ad } \mathfrak{H}) = \mathfrak{H}$ and $\mathfrak{H}$ is a Cartan subalgebra of $\mathfrak{L}$.

4.4.2.10 Proposition
A Cartan subalgebra of $\mathfrak{L}$ is a maximal nilpotent subalgebra of $\mathfrak{L}$.

PROOF. Let $\mathfrak{H}$ be a Cartan subalgebra of $\mathfrak{L}$. Let $\mathfrak{H}'$ be nilpotent and $\mathfrak{H}' \supset \mathfrak{H}$. Then $\mathfrak{H}' \subset \mathfrak{L}_0(\text{ad } \mathfrak{H}') \subset \mathfrak{L}_0(\text{ad } \mathfrak{H}) = \mathfrak{H} \subset \mathfrak{H}'$ and $\mathfrak{H}' = \mathfrak{H}$.

4.4.2.11 Proposition
Let $\mathfrak{H}$ be a subalgebra of $\mathfrak{L}$, and let k' be a field extension of k. Then $\mathfrak{H}$ is a Cartan subalgebra of $\mathfrak{L}$ iff $\mathfrak{H}_{k'}$ is a Cartan subalgebra of $\mathfrak{L}_{k'}$.

PROOF. $\mathfrak{H}$ is nilpotent iff $\mathfrak{H}_{k'}$ is nilpotent, by 2.2.3. Thus, we may assume that $\mathfrak{H}$, $\mathfrak{H}_{k'}$ are nilpotent. Now $\mathfrak{L}_0(\text{ad } \mathfrak{H})$ is a k'-form of $(\mathfrak{L}_{k'})_0(\text{ad } \mathfrak{H}_{k'})$, so that $\mathfrak{H} = \mathfrak{L}_0(\text{ad } \mathfrak{H})$ iff $\mathfrak{H}_{k'} = (\mathfrak{L}_{k'})_0(\text{ad } \mathfrak{H}_{k'})$. Thus, $\mathfrak{H}$ is a Cartan subalgebra of $\mathfrak{L}$ iff $\mathfrak{H}_{k'}$ is a Cartan subalgebra of $\mathfrak{L}_{k'}$.

4.4.3 Regular Elements and Subalgebras

4.4.3.1 Definition
A subalgebra $\mathfrak{N}$ of $\mathfrak{L}$ is *regular* in $\mathfrak{L}$ if $\mathfrak{N}$ is a nilpotent subalgebra of $\mathfrak{L}$ and $\dim \mathfrak{L}_0(\text{ad } \mathfrak{N}) \leq \dim \mathfrak{L}_0(\text{ad } \mathfrak{N}')$ for every nilpotent subalgebra $\mathfrak{N}'$ of $\mathfrak{L}$.

4.4.3.2 Definition

An element x of $\mathfrak{L}$ is *regular* if $\dim \mathfrak{L}_0(\operatorname{ad} x) \leq \dim \mathfrak{L}_0(\operatorname{ad} y)$ for all $y \in \mathfrak{L}$. The set of regular elements of $\mathfrak{L}$ is denoted $\mathfrak{L}_{\text{reg}}$.

4.4.3.3 Theorem

Let k be infinite. Then $\mathfrak{L}_{\text{reg}}$ is an open dense subset of $\mathfrak{L}$. (Here we use the Zariski topology, described in A.5.)

PROOF. Since $\mathfrak{L}_{\text{reg}}$ is nonempty, it suffices to show that it is open. Thus, let s be the maximal dimension of the spaces $\mathfrak{L}_*(\operatorname{ad} x)$ $(x \in \mathfrak{L})$, and let $n = \dim \mathfrak{L}$. Note that $\mathfrak{L}_*(\operatorname{ad} x) = \mathfrak{L}(\operatorname{ad} x)^n$, and that s is the maximal possible rank of $(\operatorname{ad} x)^n$ $(x \in \mathfrak{L})$. Furthermore, x is regular iff rank $(\operatorname{ad} x)^n = s$. Thus, x is regular iff $\det_B M(x) \neq 0$ for some $s \times s$ minor $M(x)$ of $(\operatorname{ad} x)^n$, where B is a fixed basis for $\mathfrak{L}$. The set of such x is open, since there are only finitely many such minors and since, for any fixed minor $M(x)$ of $(\operatorname{ad} x)^n$, the mapping $x \to \det_B M(x)$ is continuous so that $\{x \mid \det_B M(x) \neq 0\}$ is open.

4.4.3.4 Definition

The *rank* of $\mathfrak{L}$ is $\dim \mathfrak{L}_0(\operatorname{ad} x)$ where $x \in \mathfrak{L}_{\text{reg}}$. The rank of $\mathfrak{L}$ is denoted rank $\mathfrak{L}$.

4.4.3.5 Proposition

Let k be infinite and let $\mathfrak{N}$ be a nilpotent subalgebra of $\mathfrak{L}$. Then the following conditions are equivalent:

1. $\mathfrak{N}$ is regular;
2. $\dim \mathfrak{L}_0(\operatorname{ad} \mathfrak{N}) = \operatorname{rank} \mathfrak{L}$;
3. $\mathfrak{N}$ contains a regular element of $\mathfrak{L}$.

PROOF. We show that $1 \Rightarrow 2 \Rightarrow 3 \Rightarrow 1$. Suppose first that $\mathfrak{N}$ is regular. Take $x \in \mathfrak{L}$ such that $\mathfrak{L}_0(\operatorname{ad} \mathfrak{N}) = \mathfrak{L}_0(\operatorname{ad} x)$, by 4.3.2. Then rank $\mathfrak{L} \geq \mathfrak{L}_0(\operatorname{ad} \mathfrak{N})$ and $\dim \mathfrak{L}_0(\operatorname{ad} x) \geq \operatorname{rank} \mathfrak{L}$. Thus, $\dim \mathfrak{L}_0(\operatorname{ad} \mathfrak{N}) = \operatorname{rank} \mathfrak{L}$ and $1 \Rightarrow 2$. Suppose that $\dim \mathfrak{L}_0(\operatorname{ad} \mathfrak{N}) = \operatorname{rank} \mathfrak{L}$. Choose $x \in \mathfrak{N}$ such that $\mathfrak{L}_0(\operatorname{ad} \mathfrak{N}) = \mathfrak{L}_0(\operatorname{ad} x)$, by 4.3.2. Then $\dim \mathfrak{L}_0(\operatorname{ad} x) = \operatorname{rank} \mathfrak{L}$ and x is regular. Thus, $2 \Rightarrow 3$. Suppose, finally, that x is regular and $x \in \mathfrak{N}$. Then $\mathfrak{L}_0(\operatorname{ad} \mathfrak{N}) \subset$

$\mathfrak{L}_0(\text{ad } x)$ and $\dim \mathfrak{L}_0(\text{ad } \mathfrak{N}) \leq \dim \mathfrak{L}_0(\text{ad } x) = \text{rank } \mathfrak{L}$. Letting $\mathfrak{N}'$ be a regular subalgebra of $\mathfrak{L}$, we have $\dim \mathfrak{L}_0(\text{ad } \mathfrak{N}') \leq \dim \mathfrak{L}_0(\text{ad } \mathfrak{N}) \leq \text{rank } \mathfrak{L}$. But $\dim \mathfrak{L}_0(\text{ad } \mathfrak{N}') = \text{rank } \mathfrak{L}$, since $1 \Rightarrow 2$, so that $\dim \mathfrak{L}_0(\text{ad } \mathfrak{N}') = \dim \mathfrak{L}_0(\text{ad } \mathfrak{N})$ and $\mathfrak{N}$ is regular.

4.4.3.6 Corollary

Let k be infinite. Then x is a regular element of $\mathfrak{L}$ iff kx is a regular subalgebra of $\mathfrak{L}$.

4.4.3.7 Theorem

Let k be infinite. Then if x is a regular element of $\mathfrak{L}$, x is quasi-regular in $\mathfrak{L}$. And if $\mathfrak{H}$ is a regular subalgebra of $\mathfrak{L}$, then $\mathfrak{H}$ is quasi-regular in $\mathfrak{L}$.

PROOF. Let x be a regular element of $\mathfrak{L}$. Let $\mathfrak{N} = \mathfrak{L}_0(\text{ad } x)$. Then $\mathfrak{L} = \mathfrak{L}_0(\text{ad } x) \oplus \mathfrak{L}_*(\text{ad } x) = \mathfrak{N} \oplus \mathfrak{L}_*(\text{ad } x)$. Note that $\mathfrak{L}_*(\text{ad } x)$ is ad y-stable for $y \in \mathfrak{N}$. Let $U = \mathfrak{N}_f = \{y \in \mathfrak{N} \mid f(y) \neq 0\}$ where $f\colon \mathfrak{N} \to k$ is the polynomial function defined by $f(y) = \det(\text{ad } y|_{\mathfrak{L}_*(ad\, x)})$ for $y \in \mathfrak{N}$. Since $x \in U$, U is a nonempty Zariski open dense subset of $\mathfrak{N}$, by A.6. Now $y \in U \Rightarrow \mathfrak{L}_0(\text{ad } y) \subset \mathfrak{L}_0(\text{ad } x) \Rightarrow \mathfrak{L}_0(\text{ad } y) = \mathfrak{L}_0(\text{ad } x)$ by the minimality of $\dim \mathfrak{L}_0(\text{ad } x)$. Thus, letting $n = \dim \mathfrak{L}$, $(\text{ad } y|_{\mathfrak{N}})^n = 0$ for $y \in U$. Since $y \mapsto (\text{ad } y|_{\mathfrak{N}})^n$ is continuous, $(\text{ad } y|_{\mathfrak{N}})^n = 0$ for $y \in \bar{U} = \mathfrak{N}$. Thus, $\text{ad}_{\mathfrak{N}}\mathfrak{N}$ consists of nilpotent transformations, and $\mathfrak{N}$ is nilpotent, by 3.2.6. Thus, x is quasi-regular.

Next, let $\mathfrak{H}$ be a regular subalgebra of $\mathfrak{L}$. Then $\mathfrak{H}$ contains some regular element x of $\mathfrak{L}$, by 4.4.3.5. By the above paragraph x is quasi-regular, so $\mathfrak{L}_0(\text{ad } x)$ is nilpotent. Since $\mathfrak{L}_0(\text{ad } \mathfrak{H}) \subset \mathfrak{L}_0(\text{ad } x)$, $\mathfrak{L}_0(\text{ad } \mathfrak{H})$ is nilpotent and $\mathfrak{H}$ quasi-regular.

4.4.3.8 Corollary

Let k be infinite, x an element of $\mathfrak{L}$, and $\mathfrak{H}$ a nilpotent subalgebra of $\mathfrak{L}$. Then x (respectively $\mathfrak{H}$) is regular in $\mathfrak{L}$ iff $\mathfrak{L}_0(\text{ad } x)$ (respectively $\mathfrak{L}_0(\text{ad } \mathfrak{H})$) is a Cartan subalgebra of $\mathfrak{L}$ of minimal dimension. The minimal dimension of a Cartan subalgebra of $\mathfrak{L}$ is rank $\mathfrak{L}$.

PROOF. Let $\mathfrak{N}$ be a Cartan subalgebra of $\mathfrak{L}$. Then $\dim \mathfrak{N} = \dim \mathfrak{L}_0(\text{ad } \mathfrak{N}) \geq$

rank $\mathfrak{L}$, by 4.4.3.4. If x is regular, then $\mathfrak{L}_0(\text{ad } x)$ is a Cartan subalgebra of $\mathfrak{L}$, by 4.4.3.7 and 4.4.2.7, and dim $\mathfrak{L}_0(\text{ad } x) = \text{rank } \mathfrak{L}$, by 4.4.3.1. Thus, $\mathfrak{N}$ is a Cartan subalgebra of $\mathfrak{L}$ of minimal dimension iff dim $\mathfrak{N} = \text{rank } \mathfrak{L}$. Consequently, x is regular iff $\mathfrak{L}_0(\text{ad } x)$ is a Cartan subalgebra of $\mathfrak{L}$ of minimal dimension.

Next, let y be an element of $\mathfrak{H}$ such that $\mathfrak{L}_0(\text{ad } \mathfrak{H}) = \mathfrak{L}_0(\text{ad } y)$, by 4.3.2. Then $\mathfrak{H}$ is regular iff y is regular, by 4.4.3.5. Thus, by applying the above paragraph to $x = y$, we see that $\mathfrak{H}$ is regular iff $\mathfrak{L}_0(\text{ad } \mathfrak{H}) = \mathfrak{L}_0(\text{ad } y)$ is a Cartan subalgebra of $\mathfrak{L}$ of minimal dimension.

We will see later (and it follows easily from 3.8.8) that the dimension of a Cartan subalgebra of $\mathfrak{L}$ is constant if the characteristic of $\mathfrak{L}$ is 0. It is not always constant if the characteristic is $p > 0$. See, for instance, [26, p. 116].

4.4.3.9 Corollary

Let k be infinite. Then $\mathfrak{L}$ has a Cartan subalgebra.

PROOF. $\mathfrak{L}$ has a regular element.

If k is finite, it does not seem to be known whether every Lie algebra $\mathfrak{L}$ over k has a Cartan subalgebra. However, Der $\mathfrak{L}$ always has a Cartan subalgebra—in fact, every Lie p-algebra has a Cartan subalgebra, by 4.4.4.9. This implies that every Lie algebra has a nilpotent subalgebra of the form $\mathfrak{H} = \mathfrak{L}_0(\mathfrak{N})$ where $\mathfrak{N}$ is a nilpotent subalgebra of Der $\mathfrak{L}$ (e.g., $\mathfrak{N}$ any Cartan subalgebra of Der $\mathfrak{L}$). Such an $\mathfrak{H}$ has the property that $\mathfrak{N}_{\mathfrak{L}}(\mathfrak{H}) = \mathfrak{H} + \mathfrak{C}_{\mathfrak{L}}(\mathfrak{H})$, by 4.4.2.5. Thus, such an $\mathfrak{H}$ is a Cartan subalgebra of $\mathfrak{L}$ iff $\mathfrak{C}_{\mathfrak{L}}(\mathfrak{H}) \subset \mathfrak{H}$, by 4.4.2.9. We now consider the class of all subalgebras of the form $\mathfrak{L}_0(\mathfrak{N})$, where $\mathfrak{N}$ is a nilpotent subalgebra of Der $\mathfrak{L}$.

4.4.4 Engel Subalgebras and Fitting Subalgebras

4.4.4.1 Definition

A subalgebra of $\mathfrak{L}$ of the form $\mathfrak{L}_0(\mathfrak{N})$ where $\mathfrak{N}$ is a nilpotent subalgebra of Der $\mathfrak{L}$ is called a *Fitting subalgebra* of $\mathfrak{L}$.

4.4.4.2 Definition

A subalgebra of $\mathfrak{L}$ of the form $\mathfrak{L}_0(\mathrm{ad}\,\mathfrak{H})$ where $\mathfrak{H}$ is a nilpotent subalgebra of $\mathfrak{L}$ is called an *Engel subalgebra* of $\mathfrak{L}$.

4.4.4.3 Proposition

Let $\mathfrak{B}$ be a Fitting subalgebra of $\mathfrak{L}$. Then $\mathfrak{N}_{\mathfrak{L}}(\mathfrak{B}) = \mathfrak{B} + \mathfrak{C}_{\mathfrak{L}}(\mathfrak{B})$.

PROOF. The proof is the same as for 4.4.2.5.

4.4.4.4 Corollary

Let B be an Engel subalgebra of $\mathfrak{L}$. Then $\mathfrak{N}_{\mathfrak{L}}(\mathfrak{B}) = \mathfrak{B}$.

PROOF. By 4.4.4.3, $\mathfrak{N}_{\mathfrak{L}}(\mathfrak{B}) = \mathfrak{B} + \mathfrak{C}_{\mathfrak{L}}(\mathfrak{B})$. Now let $\mathfrak{B} = \mathfrak{L}_0(\mathrm{ad}\,\mathfrak{H})$ where $\mathfrak{H}$ is a nilpotent subalgebra of $\mathfrak{L}$. For $x \in \mathfrak{C}_{\mathfrak{L}}(\mathfrak{B})$, we have $x \in \mathfrak{C}_{\mathfrak{L}}(\mathfrak{H}) \subset \mathfrak{L}_0(\mathrm{ad}\,\mathfrak{H}) = \mathfrak{B}$ since $\mathfrak{H} \subset \mathfrak{B}$. Thus, $\mathfrak{C}_{\mathfrak{L}}(\mathfrak{B}) \subset \mathfrak{B}$ and $\mathfrak{N}_{\mathfrak{L}}(\mathfrak{B}) = \mathfrak{B}$.

We now generalize the corollary slightly.

4.4.4.5 Lemma

Let $\mathfrak{N}$ be a nilpotent Lie algebra, and let $\mathfrak{V}$, $\overline{\mathfrak{V}}$ be Lie modules for $\mathfrak{N}$ and $f\colon \mathfrak{V} \to \overline{\mathfrak{V}}$ a homomorphism from $\mathfrak{V}$ into $\overline{\mathfrak{V}}$. Then $f(\mathfrak{V}_0(\mathfrak{N})) = f(\mathfrak{V})_0(\mathfrak{N})$.

PROOF. Clearly, $f(\mathfrak{V})_0(\mathfrak{N}) \supset f(\mathfrak{V}_0(\mathfrak{N}))$ and $f(\mathfrak{V})_*(\mathfrak{N}) \supset f(\mathfrak{V}_*(\mathfrak{N}))$. But $\mathfrak{V} = \mathfrak{V}_0(\mathfrak{N}) \oplus \mathfrak{V}_*(\mathfrak{N})$ implies that $f(\mathfrak{V}) = f(\mathfrak{V}_0(\mathfrak{N})) + f(\mathfrak{V}_*(\mathfrak{N}))$. Since $f(\mathfrak{V}) = f(\mathfrak{V})_0(\mathfrak{N}) \oplus f(\mathfrak{V})_*(\mathfrak{N})$, it follows that $f(\mathfrak{V}_0(\mathfrak{N})) = f(\mathfrak{V})_0(\mathfrak{N})$ and $f(\mathfrak{V}_*(\mathfrak{N})) = f(\mathfrak{V})_*(\mathfrak{N})$.

4.4.4.6 Theorem

Let $\mathfrak{B}$ be a subalgebra of $\mathfrak{L}$ which contains an Engel subalgebra of $\mathfrak{L}$. Then $\mathfrak{N}_{\mathfrak{L}}(\mathfrak{B}) = \mathfrak{B}$.

PROOF. Let $\mathfrak{B} \supset \mathfrak{L}_0(\mathrm{ad}\,\mathfrak{H})$ where $\mathfrak{H}$ is a nilpotent subalgebra of $\mathfrak{L}$. Let $\mathfrak{N} = \mathfrak{N}_{\mathfrak{L}}(\mathfrak{B})$. Then $\mathfrak{B}$ is an ideal of $\mathfrak{N}$ and we let $f\colon \mathfrak{N} \to \mathfrak{N}/\mathfrak{B}$ be the canonical homomorphism. Regard $\mathfrak{N}$ and $f(\mathfrak{N})$ as ad $\mathfrak{H}$-modules. Then $\mathfrak{N}\,\mathrm{ad}\,\mathfrak{H} \subset \mathfrak{B} = \mathrm{Kernel}\,f$, so that $f(\mathfrak{N})\,\mathrm{ad}\,\mathfrak{H} = \{0\}$. Thus, $f(\mathfrak{N}) = f(\mathfrak{N})_0(\mathrm{ad}$

$\mathfrak{H}) = f(\mathfrak{N}_0(\text{ad } \mathfrak{H}))$ and $\mathfrak{N} \subset \mathfrak{B} + \mathfrak{N}_0(\text{ad } \mathfrak{H})$. But $\mathfrak{N}_0(\text{ad } \mathfrak{H}) \subset \mathfrak{L}_0(\text{ad } \mathfrak{H}) \subset \mathfrak{B}$. Thus, $\mathfrak{N} \subset \mathfrak{B}$ and $\mathfrak{N}_{\mathfrak{L}}(\mathfrak{B}) = \mathfrak{B}$.

4.4.4.7 Theorem

Let k be infinite. Then every Engel subalgebra of $\mathfrak{L}$ contains a Cartan subalgebra of $\mathfrak{L}$. The Cartan subalgebras of $\mathfrak{L}$ are the minimal Engel subalgebras of $\mathfrak{L}$.

PROOF. Let $\mathfrak{B} = \mathfrak{L}_0(\text{ad } \mathfrak{N})$ be an Engel subalgebra of $\mathfrak{L}$, $\mathfrak{N}$ being a nilpotent subalgebra of $\mathfrak{L}$. By 4.3.2, $\mathfrak{L}_0(\text{ad } \mathfrak{N}) = \mathfrak{L}_0(\text{ad } x)$ for some $x \in \mathfrak{N}$. Now $\mathfrak{L}_0(\text{ad } \mathfrak{N})\mathfrak{L}_*(\text{ad } x) \subset \mathfrak{L}_*(\text{ad } x)$ and the set $U = \{y \in \mathfrak{L}_0(\text{ad } \mathfrak{N}) \mid \text{ad } y|_{\mathfrak{L}_*(\text{ad } x)}$ is nonsingular$\}$ is Zariski open in $\mathfrak{B} = \mathfrak{L}_0(\text{ad } \mathfrak{N})$, by A.5. It is nonempty, since $x \in U$, hence open and dense in $\mathfrak{B}$. Thus, $U \cap \mathfrak{B}_{\text{reg}}$ is nonempty. Let $y \in U \cap \mathfrak{B}_{\text{reg}}$. Since $y \in \mathfrak{B}_{\text{reg}}$, $\mathfrak{H} = \mathfrak{B}_0(\text{ad } y)$ is a Cartan subalgebra of $\mathfrak{B}$. Since $y \in U$, $\mathfrak{L}_0(\text{ad } y) \subset \mathfrak{L}_0(\text{ad } x) = \mathfrak{B}$ and $\mathfrak{L}_0(\text{ad } y) = \mathfrak{B}_0(\text{ad } y) = \mathfrak{H}$. Thus, y is quasi-regular in $\mathfrak{L}$ and $\mathfrak{H}$ is a Cartan subalgebra of $\mathfrak{L}$, by 4.4.2.7. The remaining assertion follows directly from the first assertion.

4.4.4.8 Theorem

Let $\mathfrak{L}$ be a Lie algebra of characteristic 0 (respectively of characteristic p such that ad $\mathfrak{L}$ is closed under pth powers). Let $\mathfrak{B}$ be an Engel subalgebra of $\mathfrak{L}$. Then every Cartan subalgebra of $\mathfrak{B}$ is a Cartan subalgebra of $\mathfrak{L}$.

PROOF. Suppose first that the characteristic of $\mathfrak{L}$ is 0. We may assume, by 1.5.4 and 4.4.2.11, that k is algebraically closed. Now any Engel subalgebra $\mathfrak{B}$ of $\mathfrak{L}$ has a Cartan subalgebra $\mathfrak{H}$ of $\mathfrak{L}$, by 4.4.4.7. Let $\mathfrak{H}'$ be a second Cartan subalgebra of $\mathfrak{B}$. Then there exists, by 3.8.5, a $g \in E_{\mathfrak{L}}^{\mathfrak{B}} \subset E^{\mathfrak{L}}$ such that $\mathfrak{H}' = \mathfrak{H}g$. Thus, $\mathfrak{H}'$ is a Cartan subalgebra of $\mathfrak{L}$, since $\mathfrak{H}$ is a Cartan subalgebra of $\mathfrak{L}$ and $g \in \text{Aut } \mathfrak{L}$.

Suppose next that the characteristic of $\mathfrak{L}$ is p and that ad $\mathfrak{L}$ is closed under pth powers. Let $\mathfrak{B} = \mathfrak{L}_0(\text{ad } \mathfrak{N})$, where $\mathfrak{N}$ is a nilpotent subalgebra of $\mathfrak{L}$, be any Engel subalgebra of $\mathfrak{L}$. Let $\mathfrak{H}$ be a Cartan subalgebra of $\mathfrak{B}$. We claim that $\mathfrak{L}_0(\text{ad } \mathfrak{H}) \subset \mathfrak{L}_0(\text{ad } \mathfrak{N}) = \mathfrak{B}$, so that $\mathfrak{L}_0(\text{ad } \mathfrak{H}) = \mathfrak{B}_0(\text{ad } \mathfrak{H}) = \mathfrak{H}$ and $\mathfrak{H}$ is a Cartan subalgebra of $\mathfrak{L}$. Thus, let $x \in \mathfrak{N}$ and choose e such that $\mathfrak{B}(\text{ad } x)^{p^e} = \{0\}$. Choose $y \in \mathfrak{L}$ such that ad $y = (\text{ad } x)^{p^e}$. Since $\mathfrak{B}$ is ad x-stable, $\mathfrak{B}$ is ad y-stable and $y \in \mathfrak{N}_{\mathfrak{L}}(\mathfrak{B})$. But $\mathfrak{N}_{\mathfrak{L}}(\mathfrak{B}) = \mathfrak{B}$, by 4.4.4.4, so that $y \in \mathfrak{B}$. Since $\mathfrak{B}$ ad $y = \{0\}$, $y \in \mathfrak{C}_{\mathfrak{B}}(\mathfrak{B}) \subset \mathfrak{H}$. Thus, $\mathfrak{L}_0(\text{ad } \mathfrak{H}) \subset \mathfrak{L}_0(\text{ad } y) = \mathfrak{L}_0(\text{ad } x)$. Thus, $\mathfrak{L}_0(\text{ad } \mathfrak{H}) \subset \bigcap_{x \in \mathfrak{N}} \mathfrak{L}_0(\text{ad } x) = \mathfrak{L}_0(\text{ad } \mathfrak{N})$, and $\mathfrak{L}_0(\text{ad } \mathfrak{H}) \subset \mathfrak{L}_0(\text{ad } \mathfrak{N}) = \mathfrak{B}$.

4.4.4.9 Corollary
Let $\mathfrak{L}$ be of characteristic p and suppose that ad $\mathfrak{L}$ is closed under pth powers. Then $\mathfrak{L}$ has a Cartan subalgebra.

PROOF. The proof is by induction on dim $\mathfrak{L}$ and is trivial if dim $\mathfrak{L} = 1$ or $\mathfrak{L}$ is nilpotent. Otherwise, $\mathfrak{L}$ contains an element x such that $\mathfrak{B} = \mathfrak{L}_0(\text{ad } x)$ is not $\mathfrak{L}$. Now $\mathfrak{B}$ is an Engel subalgebra and $\mathfrak{N}_{\mathfrak{L}}(\mathfrak{B}) = \mathfrak{B}$. We claim that ad $\mathfrak{B}$ is closed under pth powers. Thus, let $x \in \mathfrak{B}$ and choose $y \in \mathfrak{L}$ such that $(\text{ad } x)^p = \text{ad } y$. Since $\mathfrak{B}(\text{ad } x)^p \subset \mathfrak{B}$, we have $y \in \mathfrak{N}_{\mathfrak{L}}(\mathfrak{B}) = \mathfrak{B}$. Thus, $(\text{ad } \mathfrak{B})^p \subset \text{ad } \mathfrak{B}$. By induction, $\mathfrak{B}$ has a Cartan subalgebra $\mathfrak{H}$. Now $\mathfrak{H}$ is a Cartan subalgebra of $\mathfrak{L}$, by 4.4.4.8.

4.4.4.10 Theorem
Let $\mathfrak{L}$ be a Lie algebra of characteristic 0 (respectively characteristic p such that ad $\mathfrak{L}$ is closed under pth powers). Then a Cartan subalgebra of a Fitting subalgebra of $\mathfrak{L}$ is quasi-regular in $\mathfrak{L}$, and is therefore contained in a unique Cartan subalgebra $\hat{\mathfrak{H}}$ of $\mathfrak{L}$.

PROOF. Let $\mathfrak{N}$ be a nilpotent subalgebra of Der $\mathfrak{L}$, $\mathfrak{H}$ a Cartan subalgebra of $\mathfrak{L}_0(\mathfrak{N})$. We are to show that $\hat{\mathfrak{H}} = \mathfrak{L}_0(\text{ad } \mathfrak{H})$ is nilpotent, for then we can invoke 4.4.2.6 to obtain our conclusions. Since the conditions on $\mathfrak{H}$ are preserved under ascent and the conclusions under descent, we may assume that k is algebraically closed. Now $\mathfrak{L} = \sum_{a \in A} \mathfrak{L}_a$ where A is the additive group of functions from $\mathfrak{N}$ to k and $\mathfrak{L}_a = \mathfrak{L}_a(\mathfrak{N})$ for $a \in A$. Since A is torsion free (respectively a p-group) and $\mathfrak{H}$ a Cartan subalgebra of $\mathfrak{L}_0$, $\mathfrak{L}_0(\text{ad } \mathfrak{H}) = \hat{\mathfrak{H}}$ is a Cartan subalgebra of $\mathfrak{L}$, by 4.3.11 and 4.4.2.6. In particular, $\mathfrak{H}$ is quasi-regular and $\hat{\mathfrak{H}}$ is the unique Cartan subalgebra of $\mathfrak{L}$ containing $\mathfrak{H}$, by 4.4.2.8.

4.4.4.11 Definition

Cart $\mathfrak{L}$ is the set of Cartan subalgebras of $\mathfrak{L}$.

The mapping $\mathfrak{H} \mapsto \hat{\mathfrak{H}}$ from the set Cart $\mathfrak{B}$ of Cartan subalgebras of a Fitting subalgebra $\mathfrak{B}$ of $\mathfrak{L}$ into Cart $\mathfrak{L}$, with $\mathfrak{L}$ as in the preceding theorem, is used in the next section in relating the Cartan subalgebras of $\mathfrak{L}$ and Cartan subalgebras of an ideal of $\mathfrak{L}$. The following representation-theoretic comparison of $\mathfrak{H}$ and $\hat{\mathfrak{H}}$ is needed for this.

4.4.4.12 Theorem

Let $\mathfrak{L}$ be a Lie algebra of characteristic 0 (respectively of characteristic p such that ad $\mathfrak{L}$ is closed under pth powers). Let $\mathfrak{N}$ be a nilpotent subalgebra of Der $\mathfrak{L}$ and let $\mathfrak{H}$ be a Cartan subalgebra of the Fitting subalgebra $\mathfrak{B} = \mathfrak{L}_0(\mathfrak{N})$. Let $\hat{\mathfrak{H}}$ be the Cartan subalgebra $\mathfrak{L}_0(\text{ad }\mathfrak{H})$ of $\mathfrak{L}$ described in 4.4.2.8. Let $\mathfrak{V}$ be a Lie module for the split extension $\mathfrak{N} \oplus \mathfrak{L}$ (respectively a Lie module for $\mathfrak{N} \oplus \mathfrak{L}$ such that $f(\mathfrak{L})$ is closed under pth powers where f is the representation of $\mathfrak{L}$ afforded by $\mathfrak{V}$). Then $\mathfrak{V}_0(\mathfrak{H}) = \mathfrak{V}_0(\hat{\mathfrak{H}})$.

PROOF. The hypothesis is preserved under ascent, the conclusion under descent. Thus, we may assume that k is algebraically closed. Choose $x \in \mathfrak{N}$ such that $\mathfrak{B} = \mathfrak{L}_0(x)$, by 4.3.2. Then $\mathfrak{B} = \mathfrak{L}_0(D)$ where $D = x_s \in \text{Der }\mathfrak{L}$, by 2.4.14. Thus, we may replace $\mathfrak{N}$ by kD. Thus, since $\mathfrak{H}D = \{0\}$, we may assume that $\mathfrak{H}\mathfrak{N} = \{0\}$. Thus, $[\text{ad }\mathfrak{H}, \mathfrak{N}] = \text{ad }(\mathfrak{H}\mathfrak{N}) = \{0\}$ and the elements of $\mathfrak{N}$ commute with the elements of ad $\mathfrak{H}$. It follows that $\hat{\mathfrak{H}} = \mathfrak{L}_0(\text{ad }\mathfrak{H})$ is $\mathfrak{N}$-stable. Also, $f(\mathfrak{L}_0(\text{ad }\mathfrak{H})) = f(\mathfrak{L})_0(\text{ad }f(\mathfrak{H}))$ is a p-subalgebra of $f(\mathfrak{L})$ if $f(\mathfrak{L})$ is closed under pth powers, by 4.2.6. By replacing $\mathfrak{L}$ by $\mathfrak{L}_0(\text{ad }\mathfrak{H})$ and $\mathfrak{N}$ by $\mathfrak{N}|_{\mathfrak{L}_0(\text{ad }\mathfrak{H})}$, we may assume that $\mathfrak{L} = \mathfrak{L}_0(\text{ad }\mathfrak{H})$ and $\hat{\mathfrak{H}} = \mathfrak{L}$. We continue to regard $\mathfrak{V}$ as $(\mathfrak{N} \oplus \mathfrak{L})$-module. Now $\mathfrak{V}_0(\mathfrak{H})$ is stable under $\mathfrak{L} = \mathfrak{L}_0(\text{ad }\mathfrak{H})$, by 3.2.13, and under $\mathfrak{N}$ since $\mathfrak{H}\mathfrak{N} = \{0\}$. Thus, $\mathfrak{V}_0(\mathfrak{H})$ is an $(\mathfrak{N} \oplus \mathfrak{L})$-submodule and we may assume without loss of generality that $\mathfrak{V} = \mathfrak{V}_0(\mathfrak{H})$. In this simplified context, what we are to prove is that $\mathfrak{V} = \mathfrak{V}_0(\mathfrak{L})$. Now $\mathfrak{L} = \sum_{a \in A} \mathfrak{L}^a$ where A is the additive group of functions from $\mathfrak{N}$ into k and $\mathfrak{L}^a = \mathfrak{L}_a(\mathfrak{N})$, $\mathfrak{V}_a = \mathfrak{V}_a(\mathfrak{N})$ for all $a \in A$. By 3.2.13, $\mathfrak{V}_b \mathfrak{L}^a \subset \mathfrak{V}_{b+a}$, so that $\mathfrak{L}$ is A-graded and $\mathfrak{V}$ is A-graded as an $\mathfrak{L}$-module. In the case that $\mathfrak{L}$ is of characteristic p and $f(\mathfrak{L})$ closed under pth powers, $\mathfrak{V}$ is a Lie p-module. Since $\mathfrak{H}$ is a Cartan subalgebra of $\mathfrak{L}^0 = \mathfrak{L}_0(\mathfrak{N})$ and $\mathfrak{L}_0(\text{ad } \mathfrak{H}) = \mathfrak{L}$, we have $\mathfrak{H} \subset \mathfrak{L}^0 \subset (\mathfrak{L}^0)_0(\text{ad } \mathfrak{H}) = \mathfrak{H}$ and $\mathfrak{H} = \mathfrak{L}^0$. Thus, $\mathfrak{B}_0(\mathfrak{L}^0) = \mathfrak{B}_0(\mathfrak{H}) = \mathfrak{B}$. Therefore $\mathfrak{B}_0(\mathfrak{L}) = \mathfrak{B}$, by 4.3.8 and 4.3.9.

4.4.5 Cartan Subalgebras of Ideals and Quotients

4.4.5.1 Theorem

Let $f: \mathfrak{L} \to \overline{\mathfrak{L}}$ be a homomorphism of Lie algebras. Then if $\mathfrak{H}$ is a Cartan subalgebra of $\mathfrak{L}$, $f(\mathfrak{H})$ is a Cartan subalgebra of $f(\mathfrak{L})$. And if $\overline{\mathfrak{H}}$ is a Cartan subalgebra of $f(\mathfrak{L})$ and $\mathfrak{H}$ a Cartan subalgebra of $f^{-1}(\overline{\mathfrak{H}})$, $\mathfrak{H}$ is a Cartan subalgebra of $\mathfrak{L}$.

PROOF. Let $\mathfrak{H}$ be a Cartan subalgebra of $\mathfrak{L}$. Regard $\mathfrak{L}$ and $\overline{\mathfrak{L}}$ as ad $\mathfrak{L}$-modules in the obvious way, so that $f: \mathfrak{L} \to \overline{\mathfrak{L}}$ is an (ad $\mathfrak{L}$)-module homomorphism. Then $f(\mathfrak{H}) = f(\mathfrak{L}_0(\text{ad } \mathfrak{H})) = f(\mathfrak{L})_0(\text{ad } \mathfrak{H}) = f(\mathfrak{L})_0(\text{ad} f(\mathfrak{H}))$ by 4.4.4.5. Thus, $f(\mathfrak{H}) = f(\mathfrak{L})_0(\text{ad} f(\mathfrak{H}))$ and $f(\mathfrak{H})$ is a Cartan subalgebra of $f(\mathfrak{L})$.

Next assume that $\overline{\mathfrak{H}}$ is a Cartan subalgebra of $f(\mathfrak{L})$ and $\mathfrak{H}$ is a Cartan subalgebra of $f^{-1}(\overline{\mathfrak{H}})$. Then $f(\mathfrak{H})$ is a Cartan subalgebra of $f(f^{-1}(\overline{\mathfrak{H}})) = \overline{\mathfrak{H}}$, by the above paragraph. Thus, $f(\mathfrak{H}) = \overline{\mathfrak{H}}$, by 4.4.2.10. We claim that $\mathfrak{H}$ is a Cartan subalgebra of $\mathfrak{L}$. Now $f(\mathfrak{L}_0(\text{ad } \mathfrak{H})) = f(\mathfrak{L})_0(\text{ad} f(\mathfrak{H})) = f(\mathfrak{L})_0(\text{ad } \overline{\mathfrak{H}}) = \overline{\mathfrak{H}}$, so $\mathfrak{L}_0(\text{ad } \mathfrak{H}) \subset f^{-1}(\overline{\mathfrak{H}})$. Thus, $\mathfrak{L}_0(\text{ad } \mathfrak{H}) = (f^{-1}(\overline{\mathfrak{H}}))_0(\text{ad } \mathfrak{H}) = \mathfrak{H}$ and $\mathfrak{H}$ is a Cartan subalgebra of $\mathfrak{L}$.

We now describe a close relationship between Cart $\mathfrak{L}$ and Cart $\mathfrak{L}'$ where $\mathfrak{L}'$ is an ideal of $\mathfrak{L}$ and where $\mathfrak{L}$ is of characteristic 0, or of characteristic p and $\text{ad}_{\mathfrak{L}'} \mathfrak{L}'$, $\text{ad}_{\mathfrak{L}} \mathfrak{L}$ are closed under pth powers. Under these conditions, we define a mapping a^* from Cart $\mathfrak{L}$ into Cart $\mathfrak{L}'$ as follows. Here, a denotes the inclusion mapping from $\mathfrak{L}'$ into $\mathfrak{L}$. For $\mathfrak{H}$ in Cart $\mathfrak{L}$, $a^*\mathfrak{H} = (\mathfrak{L}')_0(\text{ad } \mathfrak{H} \cap \mathfrak{L}')$. Note that $a^*\mathfrak{H}$ is in Cart $\mathfrak{L}'$ for $\mathfrak{H}$ in Cart $\mathfrak{L}$, since $a^*\mathfrak{H} = (\mathfrak{L}')_0(\text{ad } \mathfrak{H} \cap \mathfrak{L}') = \hat{\mathfrak{N}}$ where $\mathfrak{N}$ is the quasi-regular subalgebra $\mathfrak{N} = (\mathfrak{L}')_0(\text{ad } \mathfrak{H}) = \mathfrak{L}_0(\text{ad } \mathfrak{H}) \cap \mathfrak{L}' = \mathfrak{H} \cap \mathfrak{L}'$ of $\mathfrak{L}'$. (See 4.4.2.8). We begin by giving some alternate descriptions of $a^*(\mathfrak{H})$.

4.4.5.2 Proposition

Let $\mathfrak{L}$ be a Lie algebra, $\mathfrak{L}'$ an ideal of $\mathfrak{L}$. Suppose that $\mathfrak{L}$ is of characteristic

0 or that $\mathfrak{L}$ is of characteristic p and $\mathrm{ad}_{\mathfrak{L}'}\mathfrak{L}'$, $\mathrm{ad}_{\mathfrak{L}}\mathfrak{L}$ are closed under pth powers. Then the following conditions are equivalent for $\mathfrak{H} \in \mathrm{Cart}\,\mathfrak{L}$, $\mathfrak{H}' \in \mathrm{Cart}\,\mathfrak{L}'$:

1. $\mathfrak{H}' = a^*\mathfrak{H}$;
2. $\mathfrak{H}$ normalizes $\mathfrak{H}'$;
3. $\mathfrak{H} \subset \mathfrak{L}_0(\mathrm{ad}\,\mathfrak{H}')$;
4. $\mathfrak{H} \cap \mathfrak{L}' \subset \mathfrak{H}'$.

PROOF. If $\mathfrak{H}' = a^*\mathfrak{H}$, then $\mathfrak{H} \subset \mathfrak{L}_0(\mathrm{ad}\,(\mathfrak{H} \cap \mathfrak{L}'))$ and $\mathfrak{H}$ normalizes $\mathfrak{H}' = \mathfrak{L}_0(\mathrm{ad}(\mathfrak{H} \cap \mathfrak{L}')) \cap \mathfrak{L}'$. Thus, $1 \Rightarrow 2$. If $\mathfrak{H}$ normalizes $\mathfrak{H}'$, then obviously $\mathfrak{H} \subset \mathfrak{L}_0(\mathrm{ad}\,\mathfrak{H}')$. Thus, $2 \Rightarrow 3$. If $\mathfrak{H} \subset \mathfrak{L}_0(\mathrm{ad}\,\mathfrak{H}')$, $\mathfrak{H} \cap \mathfrak{L}' \subset \mathfrak{L}_0(\mathrm{ad}\,\mathfrak{H}') \cap \mathfrak{L}' = (\mathfrak{L}')_0(\mathrm{ad}\,\mathfrak{H}') = \mathfrak{H}'$. Thus, $3 \Rightarrow 4$. Suppose, finally, that $\mathfrak{H} \cap \mathfrak{L}' \subset \mathfrak{H}'$. Then $\mathfrak{H}' = (\mathfrak{L}')_0(\mathrm{ad}\,\mathfrak{H}') \subset (\mathfrak{L}')_0(\mathrm{ad}\,(\mathfrak{H} \cap \mathfrak{L}')) = a^*(\mathfrak{H})$. But $a^*(\mathfrak{H})$ is in $\mathrm{Cart}\,\mathfrak{L}'$ and $\mathfrak{H}'$ is, as a Cartan subalgebra of $\mathfrak{L}'$, maximal nilpotent in $\mathfrak{L}'$, by 4.4.2.10. Thus $\mathfrak{H}' = a^*(\mathfrak{H})$. Thus, $4 \Rightarrow 1$. It follows that conditions 1, 2, 3, 4 are equivalent.

4.4.5.3 Theorem

Let the hypothesis on $\mathfrak{L}$, $\mathfrak{L}'$ be as in 4.4.5.2. Then $a^*(\mathrm{Cart}\,\mathfrak{L}) = \mathrm{Cart}\,\mathfrak{L}'$ and $a^{*-1}(\mathfrak{H}') = \mathrm{Cart}\,(\mathfrak{L}_0(\mathrm{ad}\,\mathfrak{H}'))$ for $\mathfrak{H}' \in \mathrm{Cart}\,\mathfrak{L}'$.

PROOF. We know from the introductory remarks that $a^*\mathrm{Cart}\,\mathfrak{L} \subset \mathrm{Cart}\,\mathfrak{L}'$. Suppose next that $\mathfrak{H}' \in \mathrm{Cart}\,\mathfrak{L}'$. If $\mathfrak{H} \in \mathrm{Cart}\,\mathfrak{L}_0(\mathrm{ad}\,\mathfrak{H}')$, then $\mathfrak{H} \in \mathrm{Cart}\,\mathfrak{L}$, by 4.4.4.8, so that $a^*\mathfrak{H} = \mathfrak{H}'$ by 4.4.5.2 That $a^{*-1}(\mathfrak{H}') = \mathrm{Cart}\,\mathfrak{L}_0(\mathrm{ad}\,\mathfrak{H})$ now follows from 4.4.5.2. This shows, in particular, that $a^*\mathrm{Cart}\,\mathfrak{L} = \mathrm{Cart}\,\mathfrak{L}'$, since such an $\mathfrak{L}_0(\mathrm{ad}\,\mathfrak{H}')$ has a Cartan subalgebra, by 4.2.6 and 4.4.4.9.

We conclude with an alternate description of the $\mathfrak{L}_0(\mathrm{ad}\,\mathfrak{H}')$ of the above theorem.

4.4.5.4 Theorem

Let the hypothesis on $\mathfrak{L}$, $\mathfrak{L}'$ be as in 4.4.5.2. Let $\mathfrak{H} \in \mathrm{Cart}\,\mathfrak{L}$, $\mathfrak{H}' \in \mathrm{Cart}\,\mathfrak{L}$, and $a^*\mathfrak{H} = \mathfrak{H}'$. Then $\mathfrak{L}_0(\mathrm{ad}\,\mathfrak{H}') = \mathfrak{L}_0(\mathrm{ad}(\mathfrak{H} \cap \mathfrak{L}'))$.

PROOF. This is an application of 4.4.4.12. To apply 4.4.4.12, regard

$\mathfrak{V} = \mathfrak{L}$ as a Lie module for the split extension ad $\mathfrak{H} \oplus \mathfrak{L}'$. Now $\mathfrak{H} \cap \mathfrak{L}' = (\mathfrak{L}')_0(\text{ad } \mathfrak{H})$ is a Cartan subalgebra of the Fitting subalgebra $\mathfrak{V} = (\mathfrak{L}')_0(\text{ad } \mathfrak{H})$ of $\mathfrak{L}'$ (they are nilpotent and equal), so that $\mathfrak{V}_0(\mathfrak{H} \cap \mathfrak{L}') = \mathfrak{V}_0(\widehat{\mathfrak{H} \cap \mathfrak{L}'}) = \mathfrak{V}_0(\mathfrak{H}')$, by 4.4.4.12, since $\widehat{\mathfrak{H} \cap \mathfrak{L}'} = \mathfrak{L}'_0(\text{ad}(\mathfrak{H} \cap \mathfrak{L}')) = \mathfrak{H}'$. Thus, $\mathfrak{L}_0(\text{ad } \mathfrak{H}') = \mathfrak{L}_0(\text{ad}(\mathfrak{H} \cap \mathfrak{L}'))$.

4.4.6 Cartan Subalgebras and Automorphisms

The set of Cartan subalgebras of $\mathfrak{L}$ is invariant under the action of Aut $\mathfrak{L}$. If the characteristic of $\mathfrak{L}$ is 0 and k is algebraically closed, we know that Aut $\mathfrak{L}$ acts transitively on the set of Cartan subalgebras of $\mathfrak{L}$. If the characteristic of $\mathfrak{L}$ is p, the situation is much more complicated. Not only are there Lie algebras where the transitivity fails, but two Cartan subalgebras need not be of the same dimension. In the next section, we give some sufficient conditions for the invariance of the dimension of Cartan subalgebras, and of maximal tori, and a version of the transitivity problem is taken up in the case of Lie p-algebras. In the present section, we discuss conditions under which a semisimple automorphism of $\mathfrak{L}$ must stabilize some Cartan subalgebra of $\mathfrak{L}$, and consider the Cartan subalgebras of the fixed-point subalgebra of an automorphism of $\mathfrak{L}$.

4.4.6.1 Theorem

Let $g \in \text{Aut } \mathfrak{L}$ and let $\mathfrak{H}$ be a Cartan subalgebra of $\mathfrak{L}_1(g)$. Then $\mathfrak{L}_0(\text{ad } \mathfrak{H})$ is solvable.

PROOF. The hypothesis is preserved under ascent and the conclusion under descent, by 1.3 and 2.2.3. Thus, we may assume that k is algebraically closed. Let A be the multiplicative subgroup of $k^* = k - \{0\}$ generated by the eigenvalues of g. Then $\mathfrak{L} = \sum_{a \in A} \mathfrak{L}_a$ and $\mathfrak{L}_1 = \mathfrak{L}_1(g)$, where $\mathfrak{L}_a = \mathfrak{L}_a(g)$ for $a \in A$. Now A is finitely generated, so that $A = B \times F$ where B is torsion free and F finite. Since finite subgroups of k^* are cyclic, F is cyclic. Thus, A/B is cyclic and $\mathfrak{L}_0(\text{ad } \mathfrak{H})$ is solvable, by 4.3.13, since $\mathfrak{H}$ is a Cartan subalgebra of $\mathfrak{L}_1$.

4.4.6.2 Corollary

A Lie algebra having a fixed-point free automorphism is solvable.

The solvability of $\mathfrak{L}_0(\mathrm{ad}\,\mathfrak{H})$ in the above theorem enables us to reduce part of the problem of the existence of stable Cartan subalgebras of an automorphism of a Lie algebra $\mathfrak{L}$ to the special case where $\mathfrak{L}$ is solvable. We now consider this special case.

4.4.6.3 Theorem

Let $\mathfrak{L}$ be solvable, with G a finite group of semisimple automorphisms of $\mathfrak{L}$. Then $\mathfrak{L}$ has a Cartan subalgebra $\mathfrak{H}$ which is stable under G. If $\mathfrak{L}_1(G) = \{0\}$, then $\mathfrak{H}$ is unique.

PROOF. We first prove the existence of the Cartan subalgebra $\mathfrak{H}$. Suppose first that the ideal $\mathfrak{L}^\infty = \bigcap_{i=0}^{\infty} \mathfrak{L}^i$ is Abelian. By 4.4.1.1, $\mathfrak{L}$ has a Cartan subalgebra $\mathfrak{H}$ and $\mathfrak{L} = \mathfrak{H} \oplus \mathfrak{L}^\infty$. Moreover, for each $g \in G$, there exists a unique $a_g \in \mathfrak{L}^\infty$ such that $\mathfrak{H}g = \mathfrak{H} \exp \mathrm{ad}\, a_g$. The mapping $g \mapsto a_g$ on G is a 1-cocycle, i.e.,

$$a_{gh} = (a_g)h + a_h \text{ for } g, h \in G.$$

Since the order n of G is relatively prime to the characteristic of $\mathfrak{L}$ (other wise, there would be an element g of G of order $p =$ characteristic L and g would be unipotent rather than semisimple), there exists a 1-coboundary $c \in \mathfrak{L}^\infty$, i.e.,

$$a_h = c - ch \text{ for } h \in G.$$

In fact, if $b = \sum_{g \in G} a_g$, then $b - bh = \sum a_g - \sum_g (a_g)h = \sum_g a_{gh} - \sum_g (a_g)h = \sum_g a_h = n a_h$ for $h \in G$, and we take $c = \frac{1}{n} b$. For such a c, $\mathfrak{H} \exp \mathrm{ad}\, c$ is G-stable:

$$\begin{aligned}(\mathfrak{H} \exp \mathrm{ad}\, c)g &= \mathfrak{H}g\, g^{-1}(\exp \mathrm{ad}\, c)g \\ &= \mathfrak{H} \exp \mathrm{ad}\, a_g \exp \mathrm{ad}(cg) = \mathfrak{H} \exp \mathrm{ad}(c - cg) \exp \mathrm{ad}(cg) \\ &= \mathfrak{H} \exp \mathrm{ad}\, c.\end{aligned}$$

We now drop the assumption that $\mathfrak{L}^\infty$ is Abelian. We may assume that G

is finite and $\mathfrak{L}^{(1)} \neq \{0\}$. Let $\mathfrak{B}$ be the last nonzero term of the series $\mathfrak{L}^{(i)}$. Then $\mathfrak{B}$ is an ideal of $\mathfrak{L}$ stable under G and G acts on $\mathfrak{L}/\mathfrak{B}$ as a finite group of semisimple automorphisms. By an induction argument, $\mathfrak{L}$ has a G-stable subalgebra $\mathfrak{M}$ containing $\mathfrak{B}$ such that $\mathfrak{M}/\mathfrak{B}$ is a Cartan subalgebra of $\mathfrak{L}/\mathfrak{B}$. Since $\mathfrak{M}^\infty \subset \mathfrak{B}$ and $\mathfrak{B}$ is Abelian, $\mathfrak{M}^\infty$ is Abelian. Thus, $\mathfrak{M}$ has a G-stable Cartan subalgebra $\mathfrak{H}$, by the above paragraph. But then $\mathfrak{H}$ is a G-stable Cartan subalgebra of $\mathfrak{L}$, by 4.4.5.1.

The uniqueness assertion is proved by induction on dim $\mathfrak{L}$. Thus, let $\mathfrak{L}_1(G) = \{0\}$ and let $\mathfrak{H}_1$, $\mathfrak{H}_2$ be G-stable Cartan subalgebras of $\mathfrak{L}$. If dim $\mathfrak{L} \leq 1$, $\mathfrak{H}_1 = \mathfrak{H}_2$ trivially. Suppose next that $\mathfrak{L}^\infty$ is Abelian. The group $U = \exp \operatorname{ad} \mathfrak{L}^\infty$ acts simply transitively on the Cartan subalgebras of $\mathfrak{L}$, so there is a unique $u \in U$ such that $\mathfrak{H}_1^u = \mathfrak{H}_2$. Now

$$\mathfrak{H}_1^u = \mathfrak{H}_1^{gg^{-1}ug} = \mathfrak{H}_1^{g^{-1}ug},$$

so that $u = g^{-1}ug$ for $g \in G$. If $u = \exp \operatorname{ad} a$, then $a \in \mathfrak{L}_1(G) = \{0\}$, so that $u = 1$. Thus, $\mathfrak{H}_1 = \mathfrak{H}_1^1 = \mathfrak{H}_2$. Suppose, finally, that $\mathfrak{L}^\infty$ is not Abelian. Let $\mathfrak{M}$ be the last nonzero term of the series $\mathfrak{L}^{(i)}$, and let $\overline{\mathfrak{H}}_i$ be the image of $\mathfrak{H}_i$ in $\overline{\mathfrak{L}} = \mathfrak{L}/\mathfrak{M}$ under the canonical homomorphism $\mathfrak{L} \to \mathfrak{L}/\mathfrak{M} (i = 1, 2)$. The group G acts on $\overline{\mathfrak{L}}$ as a finite group of semisimple automorphisms. By Maschke's theorem (see [9]), $\overline{\mathfrak{L}}$ is G-completely reducible. Thus, $\mathfrak{L}_1(G) = \{0\}$ implies that $\overline{\mathfrak{L}}_1(G) = \{0\}$. For if $\overline{\mathfrak{L}}_1(G) = \mathfrak{N}/\mathfrak{M}$ where $\mathfrak{N}$ is a subspace of $\mathfrak{L}$ containing $\mathfrak{M}$, then $\mathfrak{N} = \mathfrak{M}' \oplus \mathfrak{M}$ with $\mathfrak{M}'$ G-stable, hence $\mathfrak{M}' \subset \mathfrak{L}_1(G) = \{0\}$ and $\mathfrak{M}' = \{0\}$. Since dim $\overline{\mathfrak{L}} <$ dim $\mathfrak{L}$, then $\overline{\mathfrak{H}}_1 = \overline{\mathfrak{H}}_2$ by induction. That is, $\mathfrak{H}_1 + \mathfrak{M} = \mathfrak{H}_2 + \mathfrak{M}$. But $\mathfrak{H}_i + \mathfrak{M}$ is a proper G-stable subalgebra of $\mathfrak{L}$, since $\mathfrak{L}^\infty$ is not Abelian. Therefore $\mathfrak{H}_1 = \mathfrak{H}_2$ by induction.

4.4.6.4 Corollary

A Lie algebra $\mathfrak{L}$ with a fixed-point-free automorphism g of finite period has a unique g-stable Cartan subalgebra.

PROOF. Since g is fixed-point-free, $\mathfrak{L}$ is solvable. If the characteristic is 0, let $h = g_s$. If the characteristic is p, let $h = g^{p^e}$ where p^e is chosen large enough that h is separable and semisimple. Then k is fixed-point free and semisimple, so that there is a unique h-stable Cartan subalgebra

$\mathfrak{H}$ of $\mathfrak{L}$, by 4.4.6.3. Since $\mathfrak{H}$ and $\mathfrak{H}g$ are h-stable, $\mathfrak{H} = \mathfrak{H}g$. Any other g-stable Cartan subalgebra $\mathfrak{H}'$ is also h-stable, so that $\mathfrak{H} = \mathfrak{H}'$ and $\mathfrak{H}$ is a unique g-stable Cartan subalgebra of $\mathfrak{L}$.

It does not seem to be known whether every completely reducible group of automorphisms of a solvable Lie algebra has a stable Cartan subalgebra, except in the case of characteristic 0. (See [20] for this case.) The above material can be sharpened somewhat, however. For details, see [32].

4.4.6.5 Definition

A group G is *supersolvable* if there is a chain $G = G_1 \supset \ldots \supset G_n = \{1\}$ of normal subgroups G_i such that G_i/G_{i+1} is cyclic for $1 \leq i \leq n-1$.

4.4.6.6 Theorem

Let $\mathfrak{L}$ be a Lie algebra of characteristic 0, or of characteristic p such that ad $\mathfrak{L}$ is closed under pth powers. Let G be a finite supersolvable group of semisimple automorphisms of $\mathfrak{L}$. Then $\mathfrak{L}$ has a Cartan subalgebra $\mathfrak{H}$ which is G-stable.

PROOF. The proof is by induction on dim $\mathfrak{L}$ and is trivial if dim $\mathfrak{L} = 1$. If $G = \{1\}$, we let $\mathfrak{H}$ be any Cartan subalgebra of $\mathfrak{L}$. Such an $\mathfrak{H}$ exists, by 4.4.4.9. Thus, let $G \neq \{1\}$ and let $\langle g \rangle$ be a nontrivial normal cyclic subgroup of G with generator g. Then $\mathfrak{L}_1(g)$ is a proper G-stable subalgebra of $\mathfrak{L}$. If the characteristic of $\mathfrak{L}$ is p and ad $\mathfrak{L}$ is closed under pth powers, then ad $\mathfrak{L}_1(g)$ is closed under pth powers. For if $x \in \mathfrak{L}_1(g)$ and $(\text{ad } x)^p =$ ad y, then ad $y =$ ad yg and ad $y =$ ad y_1, where $y = y_1 + y_*$ with $y_1 \in \mathfrak{L}_1(g)$ and $y_* \in \mathfrak{L}_*(g)$. Applying induction to $\mathfrak{L}_1(g)$ and $G|_{\mathfrak{L}_1(G)}$, $\mathfrak{L}_1(g)$ has a G-stable Cartan subalgebra $\mathfrak{H}_1$. Now $\mathfrak{L}_0(\text{ad } \mathfrak{H}_1)$ is a solvable G-stable subalgebra of $\mathfrak{L}$, and therefore has a G-stable Cartan subalgebra $\mathfrak{H}_2$. By 4.4.4.8, $\mathfrak{H}_2$ is quasi-regular, so that $\hat{\mathfrak{H}}_2 = \mathfrak{L}_0(\text{ad } \mathfrak{H}_2)$ is a G-stable Cartan subalgebra of $\mathfrak{L}$, by 4.4.2.8.

4.5 The Toral Structure of a Lie p-Algebra

We now look at a Lie p-algebra $\mathfrak{L}$ in terms of its maximal tori, defined below. The maximal tori of $\mathfrak{L}$ are closely related to the Cartan subalgebras of $\mathfrak{L}$. In fact, each Cartan subalgebra of $\mathfrak{L}$ contains a unique maximal torus of $\mathfrak{L}$ and each maximal torus of $\mathfrak{L}$ is contained in a unique Cartan subalgebra of $\mathfrak{L}$. The maximal tori of $\mathfrak{L}$ are, however, in some respects better suited than the Cartan subalgebras of $\mathfrak{L}$ as objects in terms of which to study $\mathfrak{L}$, the main reason being that some exponential techniques can be made to apply directly to maximal tori, but not to Cartan subalgebras. The reason for this is that the pth term in the exponential series is meaningless, whereas only the first two terms of the series play a role in the application to maximal tori. This will be made more explicit in 4.6.

We assume throughout 4.5 that $\mathfrak{L}$ is a finite-dimensional Lie p-algebra over k, and let K be the algebraic closure of k.

4.5.1 Definition

x^{p^e} is defined recursively, using the pth power mapping $x \mapsto x^p$, by $x^{p^e} = (x^{p^{e-1}})^p$ for $e = 2, 3, 4, \ldots$. We let $\mathfrak{L}^{p^e}$ denote the k-span of $\{x^{p^e} \mid x \in \mathfrak{L}\}$. If $x^{p^e} = 0$ for some e, x is *nilpotent*. If x is nilpotent for $x \in \mathfrak{L}$, $\mathfrak{L}$ is *nil.*

4.5.2 Definition

A *torus* of $\mathfrak{L}$ is an Abelian Lie p-subalgebra $\mathfrak{T}$ of $\mathfrak{L}$ such that $\mathfrak{T}_K$ contains no nilpotent element other than zero.

4.5.3 Definition

An element x of $\mathfrak{L}$ is *semisimple* if x is contained in some torus of $\mathfrak{L}$.

The definitions of nilpotency and semisimplicity given above are consistent with those given in 3.7.2.1, as one readily sees upon applying 4.5.5 below to the adjoint representation of $\mathfrak{L}$.

4.5.4 Proposition

An Abelian Lie p-algebra $\mathfrak{L}$ is a torus iff $\mathfrak{L} = \mathfrak{L}^p$.

PROOF. Suppose first that $\mathfrak{L}$ is a torus and let $x \in \mathfrak{L}$. We claim that $x \in \mathfrak{L}^p$. Suppose not and let j be the first integer such that x^{p^j} is a linear combination of $x, \ldots, x^{p^{j-1}}$, say

$$x^{p^j} = \sum_{i=0}^{j-1} c_i x^{p^i}.$$

Then $c_0 = 0$, since $x \notin \mathfrak{L}^p$, so that

$$x^{p^j} = (\sum_{i=1}^{j-1} d_i x^{p^{i-1}})^p$$

where the d_i are elements of K and $d_i^p = c_i$ for all i. Letting

$$y = x^{p^{j-1}} - \sum_{i=1}^{j-1} d_i^{p^{i-1}},$$

we then have $y^p = 0$, hence $y = 0$. But this is impossible, for $x, \ldots, x^{p^{j-1}}$ are linearly independent by the choice of j. Thus, $x \in \mathfrak{L}^p$ and $\mathfrak{L} = \mathfrak{L}^p$.

Suppose, conversely, that $\mathfrak{L} = \mathfrak{L}^p$. Then $\mathfrak{L}_K = \mathfrak{L}_K^p$ and $x \mapsto x^p$ must map any basis for $\mathfrak{L}_K$ to a basis for $\mathfrak{L}_K$. In particular, $x \mapsto x^p$ maps no nonzero x to zero, so that $\mathfrak{L}_K$ contains no nilpotent element and $\mathfrak{L}$ is a torus.

4.5.5 Theorem
If $\mathfrak{L}$ is a torus and f a p-representation of $\mathfrak{L}$, then $f(\mathfrak{L})$ is diagonalizable over K.

PROOF. Since we could replace $\mathfrak{L}$ by $\mathfrak{L}_K$ for this, we may assume without loss of generality that k is algebraically closed. Now $f(\mathfrak{L})$ can be put in upper triangular form, since its elements commute pairwise. Choosing p^e greater than the dimension of the vector space underlying f, each transformation x^{p^e} with $x \in f(\mathfrak{L})$ is in diagonal form. Thus, $f(\mathfrak{L})^{p^e}$ is in diagonal form. But $f(\mathfrak{L}) = f(\mathfrak{L}^{p^e}) = f(\mathfrak{L})^{p^e}$, by the preceding proposition, so that $f(\mathfrak{L})$ is diagonalizable.

4.5.6 Corollary
Let $\mathfrak{T}$ be a torus in $\mathfrak{L}$. Then $\mathfrak{L}_0(\text{ad } \mathfrak{T}) = \mathfrak{C}_{\mathfrak{L}}(\mathfrak{T})$.

PROOF. ad $\mathfrak{T}$ is diagonalizable, by 4.5.5. Therefore $\mathfrak{L}_0(\text{ad } \mathfrak{T}) = \{x \in \mathfrak{L} \mid x\text{ad } \mathfrak{T} = 0\} = \mathfrak{C}_{\mathfrak{L}}(\mathfrak{T})$.

4.5.7 Proposition

Let k be perfect and let $\mathfrak{T}$ be a torus over k. Then $\mathfrak{T} = \{x^p \mid x \in \mathfrak{T}\}$.

PROOF. Let $y \in \mathfrak{T}$. Then $y \in \mathfrak{T}^p$, by 4.5.4, so that $y = \sum_1^m \gamma_i x_i^p$ for suitable $\gamma_i \in k$ and $x_i \in \mathfrak{T}$. Choose elements $\delta_i \in k$ such that $\delta_i^p = \gamma_i$ for all i. Then $y = (\sum \delta_i x_i)^p$.

4.5.8 Theorem

Let k be perfect and $\mathfrak{L}$ Abelian. Then $\mathfrak{L} = \mathfrak{T} \oplus \mathfrak{N}$ where $\mathfrak{T}$ is a torus in $\mathfrak{L}$ and $\mathfrak{N}$ is a nil Lie p-subalgebra of $\mathfrak{L}$.

PROOF. Let $f(x) = x^p$ for $x \in \mathfrak{L}$. One verifies easily, since $k = \{\delta^{p^i} \mid \delta \in k\}$ for all i, that Kernel f^i and Image f^i are Lie p-subalgebras of $\mathfrak{L}$ for all i. Here, the f^i are viewed as homomorphisms of additive groups. Choose i sufficiently large that Kernel f^i = Kernel f^{2i} and Image f^i = Image f^{2i}. Let $\mathfrak{T}$ = Image f^i and $\mathfrak{N}$ = Kernel f^i. For $x \in \mathfrak{L}$, we can then choose $y \in \mathfrak{L}$ such that $f^i(x) = f^{2i}(y)$, that is, such that $f^i(x - f^i(y)) = 0$. But then $x - f^i(y) \in$ Kernel $f^i = \mathfrak{N}$ and $x = f^i(y) + (x - f^i(y)) \in \mathfrak{T} + \mathfrak{N}$. It follows that $\mathfrak{L} = \mathfrak{T} + \mathfrak{N}$. Now $\mathfrak{T} = \mathfrak{T}^p$ and $\mathfrak{N}^{p^i} = \{0\}$. Thus, $\mathfrak{T}$ is a torus and $\mathfrak{N}$ nil. That the sum is direct is clear, for $\mathfrak{T} \cap \mathfrak{N} = \{0\}$, by 4.5.2.

4.5.9 Corollary

Let k be perfect. Then $x = x_s + x_n$ where x_s is semisimple, x_n is nilpotent, and $x_s x_n = 0$.

PROOF. Let $\mathfrak{B}$ be the span of $x, x^p, x^{p^2}, \ldots$. Then $\mathfrak{B}$ is an Abelian Lie p-subalgebra of $\mathfrak{L}$. Thus, $\mathfrak{B} = \mathfrak{T} \oplus \mathfrak{N}$ where $\mathfrak{T}$ is a torus in $\mathfrak{B}$ and $\mathfrak{N}$ a nil Lie p-subaglebra of $\mathfrak{B}$. Now take $x = x_s + x_n$ where $x_s \in \mathfrak{T}$, $x_n \in \mathfrak{N}$.

4.5.10 Definition

For $x \in \mathfrak{L}$ (respectively $\mathfrak{S} \subset \mathfrak{L}$) we let $\langle x \rangle$ (respectively $\langle \mathfrak{S} \rangle$) be the intersection of all Lie p-subalgebras of $\mathfrak{L}$ containing x (respectively $\mathfrak{S}$).

Obviously, $\langle x \rangle$ is the span of $x, x^p, x^{p^2}, \ldots$ and $\langle x \rangle$ is Abelian. If the elements of $\mathfrak{S}$ commute pairwise, then $\mathfrak{S} = \sum_{s \in S} \langle s \rangle$ and $\mathfrak{S}$ is Abelian.

4.5.11 Proposition

Let x be a semisimple element of $\mathfrak{L}$. Then $\langle x \rangle = \langle x^p \rangle$. Suppose that the elements of the set $\mathfrak{S}$ are semisimple elements of $\mathfrak{L}$ which commute pairwise. Then $\langle \mathfrak{S} \rangle$ is a torus.

PROOF. Let $x \in \mathfrak{T}$, $\mathfrak{T}$ being a torus of $\mathfrak{L}$. Then $\langle x \rangle \subset \mathfrak{T}$, so that $\langle x \rangle$ is a torus. Thus, $\langle x \rangle = \langle x \rangle^p = \langle x^p \rangle$, by 4.5.4. In particular, $\langle s \rangle$ is a torus for $s \in \mathfrak{S}$. Choosing $s_1, \ldots, s_m \in \mathfrak{S}$ such that $\langle \mathfrak{S} \rangle = \sum_1^m \langle s_1 \rangle$, we have $\langle \mathfrak{S} \rangle^p = \sum \langle s_i \rangle^p = \sum \langle s_i \rangle = \langle \mathfrak{S} \rangle$, and $\langle \mathfrak{S} \rangle$ is a torus, by 4.5.4.

4.5.12 Corollary

If x is a semisimple (respectively nilpotent) element of $\mathfrak{L}$ and f a p-representation of $\mathfrak{L}$, then $f(x)$ is semisimple (respectively nilpotent).

PROOF. If x is semisimple, then $\langle x \rangle = \langle x^p \rangle$. Thus, $\langle f(x) \rangle = \langle f(x)^p \rangle = \langle f(x) \rangle^p$ and $\langle f(x) \rangle$ is a torus. Thus, $f(x)$ is semisimple. If x is nilpotent, $x^{p^e} = 0$, hence $f(x)^{p^e} = 0$, for some e. Thus, if x is nilpotent, so is $f(x)$.

4.5.13 Corollary

If $\mathfrak{L}$ is a torus (respectively $\mathfrak{L}$ is nil) and f a p-representation of $\mathfrak{L}$, then $f(\mathfrak{L})$ is a torus (respectively $f(\mathfrak{L})$ is nil).

PROOF. Use the preceding corollary and proposition.

4.5.14 Proposition

Let $\mathfrak{T}$ be an ideal of $\mathfrak{L}$ such that $\mathfrak{T}$ and $\mathfrak{L}/\mathfrak{T}$ are tori. Then $\mathfrak{L}$ is a torus.

PROOF. Let $\mathfrak{S}$ be a maximal Abelian subalgebra of $\mathfrak{L}$ containing $\mathfrak{T}$. Then $\langle \mathfrak{S} \rangle$ is Abelian, so $\mathfrak{S} = \langle \mathfrak{S} \rangle$ and $\mathfrak{S}$ is a p-subalgebra of $\mathfrak{L}$. Now $(\mathfrak{S}/\mathfrak{T})^p = \mathfrak{S}/\mathfrak{T}$, since $\mathfrak{S}/\mathfrak{T}$ is a torus, so that $\mathfrak{S}^p = \mathfrak{S}^p + \mathfrak{T}^p = \mathfrak{S}^p + \mathfrak{T} = \mathfrak{S}$. Thus, $\mathfrak{S}$ is a torus. Now $\mathfrak{S}$ is an ideal of $\mathfrak{L}$, since $\mathfrak{L}/\mathfrak{T}$ is Abelian. Therefore $\mathfrak{L} = \mathfrak{L}_0(\operatorname{ad} \mathfrak{S}) = \mathfrak{C}_{\mathfrak{L}}(\mathfrak{S})$, by 4.5.6. But then $\mathfrak{S}$ is both central and maximal Abelian, so that $\mathfrak{S} = \mathfrak{L}$. Thus, $\mathfrak{L}$ is a torus.

4.5.15 Corollary

Let $\mathfrak{L}$ be nilpotent. Then $\mathfrak{L}$ has a unique maximal torus $\mathfrak{T}$. Furthermore,

$\mathfrak{T}$ is central in $\mathfrak{L}$, $\mathfrak{L}/\mathfrak{T}$ is a nil Lie p-algebra and $\mathfrak{V}_0(\mathfrak{L}) = \mathfrak{V}_0(\mathfrak{T})$ for every finite-dimensional Lie p-module $\mathfrak{V}$ for $\mathfrak{L}$.

PROOF. Let $\mathfrak{T}$ be a maximal torus of $\mathfrak{L}$. Since $\mathfrak{L}$ is nilpotent, $\mathfrak{L} = \mathfrak{L}_0(\text{ad } \mathfrak{T}) = \mathfrak{C}_{\mathfrak{L}}(\mathfrak{T})$ and $\mathfrak{T}$ is central in $\mathfrak{L}$. If $\mathfrak{S}$ is a maximal torus of $\mathfrak{L}$, then $\mathfrak{S} + \mathfrak{T}$ is a torus of $\mathfrak{L}$, by 4.5.11, so that $\mathfrak{S} = \mathfrak{S} + \mathfrak{T} = \mathfrak{T}$. We claim next that $\mathfrak{L}/\mathfrak{T}$ is nil. Suppose that $x \in \mathfrak{L}$ and let $\mathfrak{B}$ be the Abelian p-subalgebra $\langle x \rangle + \mathfrak{T}$. Choose e such that $\mathfrak{B}^{pe} = \mathfrak{B}^{pe+1}$. Then $(\mathfrak{B}^{pe})^p = \mathfrak{B}^{pe}$, so that $\mathfrak{B}^{pe}$ is a torus containing $\mathfrak{T}$. Thus, $\mathfrak{B}^{pe} = \mathfrak{T}$ and $x^{pe} \in \mathfrak{T}$. It follows that every element $x + \mathfrak{T}$ of $\mathfrak{L}/\mathfrak{T}$ is nilpotent, thus that $\mathfrak{L}/\mathfrak{T}$ is nil. Finally, let $\mathfrak{V}$ be a finite-dimensional Lie p-module for $\mathfrak{L}$. Then $\mathfrak{V}_0(\mathfrak{L}) \subset \mathfrak{V}_0(\mathfrak{T})$, and $\mathfrak{V}_0(\mathfrak{T})$ is $\mathfrak{L}$-stable since $\mathfrak{T}$ is central. Letting $x \in \mathfrak{L}$, we have $x^{pe} \in \mathfrak{T}$ for some e, and the transformation of $\mathfrak{V}_0(\mathfrak{T})$ induced by x^{pe}, hence the transformation of $\mathfrak{V}_0(\mathfrak{T})$ induced by x, is nilpotent. It follows that $\mathfrak{V}_0(\mathfrak{L}) = \mathfrak{V}_0(\mathfrak{T})$.

4.5.16 Corollary

For $x \in \mathfrak{L}$, x^{pe} is semisimple for some e.

PROOF. Let $\mathfrak{T}$ be a maximal torus of $\langle x \rangle$. Then $\langle x \rangle/\mathfrak{T}$ is nil, so $x^{pe} \in \mathfrak{T}$ for some e.

4.5.17 Theorem

$\mathfrak{H}$ is a Cartan subalgebra of $\mathfrak{L}$ iff $\mathfrak{H} = \mathfrak{C}_{\mathfrak{L}}(\mathfrak{T})$ for some maximal torus $\mathfrak{T}$ of $\mathfrak{L}$.

PROOF. Suppose first that $\mathfrak{H}$ is a Cartan subalgebra of $\mathfrak{L}$. If $x \in \mathfrak{H}$, then $\mathfrak{H}$ is stable under ad x, hence under $(\text{ad } x)^p = \text{ad } x^p$, so that $x^p \in \mathfrak{N}_{\mathfrak{L}}(\mathfrak{H}) = \mathfrak{H}$. Thus, $\mathfrak{H}$ is a Lie p-subalgebra of $\mathfrak{L}$. Let $\mathfrak{T}$ be the maximal torus of $\mathfrak{H}$, so that $\mathfrak{H} = \mathfrak{L}_0(\text{ad } \mathfrak{H}) = \mathfrak{L}_0(\text{ad } \mathfrak{T}) = \mathfrak{C}_{\mathfrak{L}}(\mathfrak{T})$, by 4.5.6 and 4.5.15. If $\mathfrak{S}$ is a maximal torus of $\mathfrak{L}$ containing $\mathfrak{T}$, then $\mathfrak{S} \subset \mathfrak{C}_{\mathfrak{L}}(\mathfrak{T}) = \mathfrak{H}$, so that $\mathfrak{S} = \mathfrak{T}$, by 4.5.15. Thus, $\mathfrak{T}$ is a maximal torus of $\mathfrak{L}$. Suppose, conversely, that $\mathfrak{T}$ is a maximal torus of $\mathfrak{L}$ and $\mathfrak{H} = \mathfrak{C}_{\mathfrak{L}}(\mathfrak{T})$. Let $x \in \mathfrak{H}$ and choose e such that x^{pe} is semisimple. Since $x^{pe} \in \mathfrak{C}_{\mathfrak{L}}(\mathfrak{T})$, $\mathfrak{T} + \langle x^{pe} \rangle$ is a torus, by 4.5.10. Since $\mathfrak{T}$ is a maximal torus, $x^{pe} \in \mathfrak{T}$. That is, $\mathfrak{H}/\mathfrak{T}$ is a nil p-subalgebra of

$\mathfrak{L}$. By 3.2.6, $\mathfrak{H}/\mathfrak{T}$ is nilpotent. Since $\mathfrak{T}$ is central, it follows that $\mathfrak{H}$ is nilpotent. Finally, $\mathfrak{H} = \mathfrak{C}_{\mathfrak{L}}(\mathfrak{T}) = \mathfrak{L}_0(\text{ad } \mathfrak{T}) \supset \mathfrak{L}_0(\text{ad } \mathfrak{H}) \supset \mathfrak{H}$, so that $\mathfrak{H} = \mathfrak{L}_0(\text{ad } \mathfrak{H})$ and $\mathfrak{H}$ is a Cartan subalgebra of $\mathfrak{L}$.

4.5.18 Theorem

Let E be an extension field of k, and let $\mathfrak{T}$ be a maximal torus of $\mathfrak{L}$. Then $\mathfrak{T}_E$ is a maximal torus of $\mathfrak{L}_E$.

PROOF. Since $\mathfrak{T}$ is a maximal torus of $\mathfrak{L}$, $\mathfrak{H} = \mathfrak{C}_{\mathfrak{L}}(\mathfrak{T})$ is a Cartan subalgebra of $\mathfrak{L}$. It follows that $\mathfrak{H}_E$ is a Cartan subalgebra of $\mathfrak{L}_E$, by 4.4.2.11, hence that $\mathfrak{H}_E$ contains a unique maximal torus of $\mathfrak{L}_E$, by 4.5.17. Consequently, we may assume without loss of generality that $\mathfrak{L}$ is nilpotent. Now $\mathfrak{T} \subset \mathfrak{C} = \mathfrak{C}(\mathfrak{L})$, by 4.5.15, and $\mathfrak{T}_E \subset \mathfrak{C}_E$. One verifies easily, using 1.3, that $\mathfrak{C}_E$ is the center of $\mathfrak{L}_E$. A maximal torus of $\mathfrak{L}_E$ is contained in $\mathfrak{C}_E$, by 4.5.15. Thus, we could replace $\mathfrak{L}$ by $\mathfrak{C}$, that is, we may assume without loss of generality that $\mathfrak{L}$ is Abelian. Now choose e such that $\mathfrak{L}^{p^e} = \mathfrak{L}^{p^{e+1}}$. Then $\mathfrak{L}^{p^e} = (\mathfrak{L}^{p^e})^p$, so that $\mathfrak{L}^{p^e}$ is a torus containing $\mathfrak{T}$. Thus, $\mathfrak{T} = \mathfrak{L}^{p^e}$. It follows that $\mathfrak{T}_E = \mathfrak{L}_E^{p^e}$. Now $\mathfrak{T}_E$ is a torus, since $\mathfrak{T}_E = (\mathfrak{T}_E)^p$, and it must be a maximal torus, since $\mathfrak{T}_E = (\mathfrak{L}_E)^{p^e}$.

4.5.19 Corollary

If $\mathfrak{L}$ is a torus and E an extension field of k, then $\mathfrak{L}_E$ is a torus.

4.5.20 Theorem

Let $0 \to \mathfrak{L}' \xrightarrow{f} \mathfrak{L} \xrightarrow{g} \overline{\mathfrak{L}} \to 0$ be an exact sequence of Lie p-algebras, the f, g being p-homomorphisms. Then if $\mathfrak{T}$ is a maximal torus of $\mathfrak{L}$, $g(\mathfrak{T})$ is a maximal torus of $\overline{\mathfrak{L}}$. And if $\overline{\mathfrak{T}}$ is a maximal torus of $\overline{\mathfrak{L}}$, any maximal torus $\mathfrak{T}$ of $g^{-1}(\overline{\mathfrak{T}})$ is a maximal torus of $\mathfrak{L}$ such that $g(\mathfrak{T}) = \overline{\mathfrak{T}}$.

PROOF. Let $\mathfrak{T}$ be a maximal torus of $\mathfrak{L}$. Then $\mathfrak{H} = \mathfrak{C}_{\mathfrak{L}}(\mathfrak{T})$ is a Cartan subalgebra of $\mathfrak{L}$, by 4.5.17, so that $g(\mathfrak{H})$ is a Cartan subalgebra of $g(\mathfrak{L}) = \overline{\mathfrak{L}}$, by 4.4.5. Now $\mathfrak{H}/\mathfrak{T}$ is nil, so that $g(\mathfrak{H})/g(\mathfrak{T})$ is nil. Thus, $g(\mathfrak{T})$ is the maximal torus of $g(\mathfrak{H})$. It follows that $g(\mathfrak{T})$ is a maximal torus of $\mathfrak{L}$, by 4.5.17.

Conversely, let $\overline{\mathfrak{T}}$ be a maximal torus of $\overline{\mathfrak{L}}$, $\mathfrak{T}$ a maximal torus of $g^{-1}(\overline{\mathfrak{T}})$. Let $\mathfrak{S}$ be a maximal torus of $\mathfrak{L}$ containing $\mathfrak{T}$. Then $g(\mathfrak{S})$ is a maximal torus of $\overline{\mathfrak{L}}$ containing $\overline{\mathfrak{T}}$, by the first paragraph of this proof. Thus, $g(\mathfrak{S}) = \overline{\mathfrak{T}}$ and $\mathfrak{S} \subset g^{-1}(\overline{\mathfrak{T}})$. Consequently, $\mathfrak{S} = \mathfrak{T}$ and $\mathfrak{T}$ is a maximal torus of $\mathfrak{L}$ such that $g(\mathfrak{T}) = \overline{\mathfrak{T}}$.

4.6 Exponentials

In 3.8, exponentials were used in showing that any two Cartan subalgebras of a Lie algebra $\mathfrak{L}$ over an algebraically closed field of characteristic 0 were conjugate under an inner automorphism of $\mathfrak{L}$. In this section, we discuss a substitute for exponentials which is used in exploring the distribution of the maximal tori in a Lie p-algebra $\mathfrak{L}$. The discussion is somewhat sketchy, and we refer the reader to [34] for a more detailed account.

Throughout the section, $\mathfrak{L}$ is a finite-dimensional Lie p-algebra over an algebraically closed field k of characteristic p.

4.6.1 Definition

$$E^x = \sum_0^{p-1} \frac{(\text{ad } x)^n}{n!} \quad \text{for } x \in \mathfrak{L}.$$

If ad x is nilpotent, E^x is nonsingular. However, E^x need not be an automorphism of $\mathfrak{L}$ even if $(\text{ad } x)^p = 0$.

Let $\mathfrak{T}$ be a torus, x an element of $\mathfrak{L}_a(\mathfrak{T})$, a being a k-valued function on $\mathfrak{T}$. Let $\mathfrak{B}$ be the Lie p-algebra $\langle \mathfrak{T} \cup \{x\} \rangle$ generated by $\mathfrak{T}$ and x. Then $\mathfrak{B} = \mathfrak{T} + \mathfrak{A}$ (not necessarily direct), where $\mathfrak{A} = \langle x \rangle$ and $\mathfrak{A}$ is an Abelian p-ideal of $\mathfrak{B}$. Since $x \in \mathfrak{A}$ and $\mathfrak{A}$ is an Abelian ideal of $\mathfrak{B}$, $E^x|_{\mathfrak{B}}$ is an automorphism of $\mathfrak{B}$ and $\mathfrak{T}E^x$ is an Abelian subalgebra of $\mathfrak{B}$.

4.6.2 Definition

In the context above, $\mathfrak{T}e^x$ is the maximal torus of the Abelian p-subalgebra $\langle \mathfrak{T}E^x \rangle$ of $\mathfrak{L}$.

4.6.3 Theorem

Let $\mathfrak{T}$ be a maximal torus of $\mathfrak{L}$, $x \in \mathfrak{L}_a(\mathfrak{T})$ for some a. Then $\dim \mathfrak{T} = \dim \mathfrak{T}e^x$.

PROOF. Let $\mathfrak{A}$, $\mathfrak{B}$ be as above. Since $\mathfrak{T}$, x, $\mathfrak{T}e^x$ are contained in $\mathfrak{B}$, we may restrict our attention to $\mathfrak{B}$. Since $(\text{ad } x|_{\mathfrak{B}})^2 = 0$, $E^x|_{\mathfrak{B}}$ is an automorphism of $\mathfrak{B}$. Let $f = E^x|_{\mathfrak{B}}$. Then the diagram

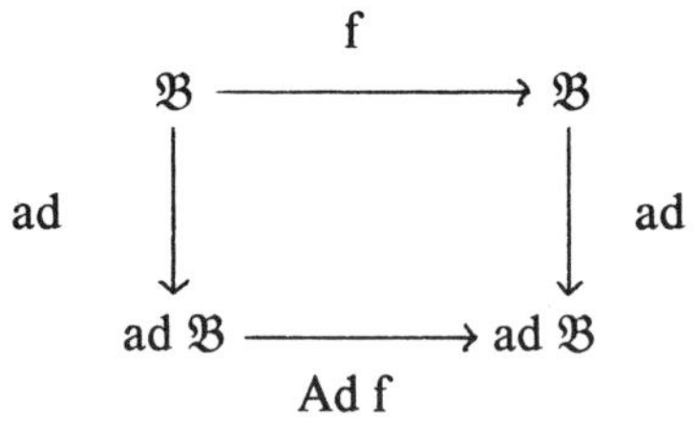

Figure 36.

is commutative, where $\text{ad } b \text{ Ad} f = f^{-1}\text{ad } bf$ for $\text{ad } b \in \text{ad } \mathfrak{B}$. Clearly, ad is a p-homomorphism, and Ad f is a p-automorphism of ad $\mathfrak{B}$ (although f need not be a p-automorphism of $\mathfrak{B}$). Thus, $(\text{ad } \mathfrak{T})\text{Ad} f = \text{ad } f(\mathfrak{T}) = \text{ad } \mathfrak{T}E^x$ is a maximal torus of ad $\mathfrak{B}$ and $\dim \text{ad } \mathfrak{T} = \dim \text{ad } \mathfrak{T}E^x$. Now $\mathfrak{T}e^x$ is a maximal torus of $\langle \mathfrak{T}E^x \rangle$, and $\mathfrak{T}e^x$ contains the maximal torus $\mathfrak{S}$ of $\mathfrak{C} = \mathfrak{C}_{\mathfrak{B}}(\mathfrak{B}) = \text{Kernel ad}|_{\mathfrak{B}}$, since $\mathfrak{T}$ does, so that $\mathfrak{T}e^x$ is a maximal torus of $\langle \mathfrak{T}E^x \rangle + \mathfrak{C} = \text{ad}^{-1}(\text{ad } \mathfrak{T}E^x)$. Thus, ad $\mathfrak{T}e^x$ is a maximal torus of ad $\mathfrak{T}E^x$, by 4.5.20, that is, $\text{ad } \mathfrak{T}e^x = \text{ad } \mathfrak{T}E^x$. It follows that $\dim \text{ad } \mathfrak{T} = \dim \text{ad } \mathfrak{T}e^x$. Since $\mathfrak{S}$ is a maximal torus of $\mathfrak{C} = \text{Kernel ad}|_{\mathfrak{B}}$ and $\mathfrak{S}$ is contained in $\mathfrak{T}$ and $\mathfrak{T}e^x$, we have $\mathfrak{S} = \mathfrak{C} \cap \mathfrak{T} = \mathfrak{C} \cap \mathfrak{T}e^x$ and $\dim \mathfrak{T} = \dim \mathfrak{S} + \dim \text{ad } \mathfrak{T} = \dim \mathfrak{S} + \dim \text{ad } \mathfrak{T}e^x = \dim \mathfrak{T}e^x$. Thus, $\dim \mathfrak{T} = \dim \mathfrak{T}e^x$.

4.6.4 Theorem

Let $\mathfrak{T}$ be a maximal torus of $\mathfrak{L}$, $x \in \mathfrak{L}_a(\mathfrak{T})$. Then $\mathfrak{T}$ is a Cartan subalgebra of $\mathfrak{L}$ iff $\mathfrak{T}e^x$ is a Cartan subalgebra of $\mathfrak{L}$.

PROOF. We may take $a \neq 0$, for otherwise $\mathfrak{T} = \mathfrak{T}e^x$. Suppose first that $\mathfrak{T}$ is a Cartan subalgebra of $\mathfrak{L}$. We claim that $\mathfrak{T}e^x = \mathfrak{C}_{\mathfrak{L}}(\mathfrak{T}e^x)$, hence that $\mathfrak{T}e^x$ is a Cartan subalgebra of $\mathfrak{L}$, by 4.5.17. Let $\mathfrak{T}_0$ be the kernel of a in $\mathfrak{T}$. Then $\mathfrak{C}_{\mathfrak{L}}(\mathfrak{T}_0)$ contains $\mathfrak{T}$, x, $\mathfrak{T}e^x$ and $\mathfrak{C}_{\mathfrak{L}}(\mathfrak{T}e^x)$. To show that $\mathfrak{T}e^x = \mathfrak{C}_{\mathfrak{L}}(\mathfrak{T}e^x)$, we may therefore assume that $\mathfrak{L} = \mathfrak{C}_{\mathfrak{L}}(\mathfrak{T}_0)$, or that $\mathfrak{T}_0$ is central in $\mathfrak{L}$. Now $0 = \mathfrak{T}(\text{ad } x)^p = \mathfrak{T} \text{ ad } x^p$, so $x^p \in \mathfrak{C}_{\mathfrak{L}}(\mathfrak{T}) = \mathfrak{T}$. It follows that $x^p \in \mathfrak{T}_0$, since $0 = x \text{ ad } x^p = a(x^p)x$. Thus, $\mathfrak{L} \text{ ad } x^p = \mathfrak{L}(\text{ad } x)^p = 0$ and $(\text{ad } x)^p = 0$. Now, a straightforward calculation, based on $(\text{ad } x)^p = 0$, shows that $\mathfrak{C}_{\mathfrak{L}}(\mathfrak{T}E^x) = \mathfrak{C}_{\mathfrak{L}}(\mathfrak{T})E^x = \mathfrak{T}E^x$ and that ad $\mathfrak{T}E^x$ is diagonalizable. It follows from 4.7.15 that $\mathfrak{C}_{\mathfrak{L}}(\mathfrak{T}e^x) = \mathfrak{C}_{\mathfrak{L}}(\mathfrak{T}E^x) = \mathfrak{T}E^x$. But then $\mathfrak{T}e^x \subset \mathfrak{T}E^x$ and, since they are of the same dimension, $\mathfrak{T}e^x = \mathfrak{T}E^x$. Thus, $\mathfrak{C}_{\mathfrak{L}}(\mathfrak{T}e^x) = \mathfrak{T}e^x$.

Suppose, conversely, that $\mathfrak{T}e^x$ is a Cartan subalgebra of $\mathfrak{L}$. Then one easily shows that $\mathfrak{T} = \mathfrak{T}e^x e^{-x}$, hence that $\mathfrak{T}$ is a Cartan subalgebra of $\mathfrak{L}$, by the above paragraph.

4.6.5 Definition

Let $\mathfrak{H}$ be a Cartan subalgebra of $\mathfrak{L}$. Then $\mathfrak{H}e^x = \mathfrak{C}_{\mathfrak{L}}(\mathfrak{T}e^x)$ where $\mathfrak{T}$ is the maximal torus of $\mathfrak{H}$, provided that $\mathfrak{T}e^x$ is defined.

4.6.6 Definition

The *rank* of a Cartan subalgebra $\mathfrak{H}$ of $\mathfrak{L}$ is the dimension of its maximal torus.

By 4.5.17, the Cartan subalgebras of maximal rank are those that contain maximal tori of maximal dimension.

4.6.7 Proposition

Let $\mathfrak{H}$ be a Cartan subalgebra of $\mathfrak{L}$ of maximal rank. Then $\mathfrak{H}e^x$ is a Cartan subalgebra of $\mathfrak{L}$ of maximal rank.

PROOF. Let $\mathfrak{T}$ be the maximal torus of $\mathfrak{H}$. Then $\dim \mathfrak{T} = \dim \mathfrak{T}e^x$, so $\mathfrak{T}e^x$ is a torus of $\mathfrak{L}$ of maximal dimension. Thus, $\mathfrak{C}_{\mathfrak{L}}(\mathfrak{T}e^x)$ is a Cartan subalgebra of $\mathfrak{L}$ of maximal rank.

Note in the above that if $\mathfrak{H}$ is a Cartan subalgebra not of maximal rank, there is given no assurance that $\mathfrak{H}e^x$ is a Cartan subalgebra of $\mathfrak{L}$.

4.6.8 Theorem

Let $\mathfrak{L}$ be solvable, and let $\mathfrak{T}, \mathfrak{T}'$ (respectively $\mathfrak{H}, \mathfrak{H}'$) be two maximal tori (respectively Cartan subalgebras) of $\mathfrak{L}$. Then there exist $x_1, \ldots, x_n \in \mathfrak{L}^\infty$ and maximal tori $\mathfrak{T}_1, \ldots, \mathfrak{T}_n$ (respectively Cartan subalgebras $\mathfrak{H}_1, \ldots, \mathfrak{H}_n$) of $\mathfrak{L}$ such that $\mathfrak{T}_i e^{x_i}$ (respectively $\mathfrak{H}_i e^{x_i}$) is defined and is

$\mathfrak{T}_{i+1}$ (respectively $\mathfrak{H}_{i+1}$) for $1 \leq i \leq n-1$ and such that $\mathfrak{T} = \mathfrak{T}_1$, $\mathfrak{T}_n = \mathfrak{T}'$ (respectively $\mathfrak{H} = \mathfrak{H}_1$, $\mathfrak{H}_n = \mathfrak{H}'$).

PROOF. By 4.5.17, it suffices to prove the theorem for the tori $\mathfrak{T}$, $\mathfrak{T}'$. If we can find a maximal torus to which all other maximal tori are conjugate in the above sense, then any two such $\mathfrak{T}$, $\mathfrak{T}'$ are also conjugate in that sense. Thus, we may specify $\mathfrak{T}$. It is convenient to choose $\mathfrak{T}$ to be a torus of maximal dimension. We now proceed by induction on $\dim \mathfrak{L}$, the case $\dim \mathfrak{L} = 1$ being trivial. Let $\mathfrak{A}$ be a minimal nonzero Abelian p-ideal of $\mathfrak{L}$. Then $\mathfrak{A}^p = \mathfrak{A}$ or $\mathfrak{A}^p = \{0\}$, so either $\mathfrak{A}$ is a torus or $\mathfrak{A}$ is nil, by 4.5.4. Let $f: \mathfrak{L} \to \mathfrak{L}/\mathfrak{A}$ be the canonical p-homomorphism. Then $f(\mathfrak{T})$, $f(\mathfrak{T}')$ are maximal tori of $f(\mathfrak{L})$, by 4.5.20, so that there exist $\bar{x}_1, \ldots, \bar{x}_n \in f(\mathfrak{L})$ and maximal tori $\overline{\mathfrak{T}}_1, \ldots, \overline{\mathfrak{T}}_n$ of $f(\mathfrak{L})$ such that $\overline{\mathfrak{T}}_i e^{\bar{x}_i}$ is defined and is $\overline{\mathfrak{T}}_{i+1}$ for $1 \leq i \leq n-1$ and $f(\mathfrak{T}) = \overline{\mathfrak{T}}_1$, $\overline{\mathfrak{T}}_n = f(\mathfrak{T}')$. Letting $\mathfrak{T}_i$ be a maximal torus of $f^{-1}(\overline{\mathfrak{T}}_i)$, we have $f(\mathfrak{T}_i) = \overline{\mathfrak{T}}_i$, by 4.5.20, and it follows that there exists an $x_i \in f^{-1}(\bar{x}_i)$ such that $x_i \in \mathfrak{L}_{a_i}(\mathfrak{T}_i)$ for some a_i. One easily sees that $f(\mathfrak{T}_i e^{x_i}) = f(\mathfrak{T}_i) e^{f(x_i)} = \overline{\mathfrak{T}}_i e^{\bar{x}_i} = \overline{\mathfrak{T}}_{i+1} = f(\mathfrak{T}_{i+1})$ for $1 \leq i \leq n-1$. Also, $f(\mathfrak{T}) = f(\mathfrak{T}_1)$ and $f(\mathfrak{T}_n) = f(\mathfrak{T}')$. It follows that $\mathfrak{T} + \mathfrak{A} = \mathfrak{T}_1 + \mathfrak{A}$, $\mathfrak{T}_i e^{x_i} + \mathfrak{A} = \mathfrak{T}_{i+1} + \mathfrak{A}$ for $1 \leq i \leq n-1$ and $\mathfrak{T}_n + \mathfrak{A} = \mathfrak{T}' + \mathfrak{A}$. If $\mathfrak{A}$ is a torus, it is central, by 4.5.6, hence contained in $\mathfrak{T}$, in the $\mathfrak{T}_i$, and in $\mathfrak{T}'$. In this case, there is nothing more to prove, since $\mathfrak{T} = \mathfrak{T}_1$, $\mathfrak{T}_i e^{x_i} = \mathfrak{T}_{i+1}$, $\mathfrak{T}_n = \mathfrak{T}'$. Assume next that $\mathfrak{A}$ is not a torus, so that $\mathfrak{A}^p = \{0\}$. It suffices to show, using $\mathfrak{T} + \mathfrak{A} = \mathfrak{T}_1 + \mathfrak{A}$, $\mathfrak{T}_i e^{x_i} + \mathfrak{A} = \mathfrak{T}_{i+1} + \mathfrak{A}$ for $1 \leq i \leq n-1$, $\mathfrak{T}_n + \mathfrak{A} = \mathfrak{T}' + \mathfrak{A}$, that $\mathfrak{T}$ is conjugate to $\mathfrak{T}_1$, $\mathfrak{T}_i e^{x_i}$ is conjugate to $\mathfrak{T}_{i+1}$ for $1 \leq i \leq n-1$, and $\mathfrak{T}_n$ is conjugate to $\mathfrak{T}' + \mathfrak{A}$. For this, we may assume without loss of generality that $\mathfrak{T} = \mathfrak{A} = \mathfrak{T}' + \mathfrak{A} = \mathfrak{L}$ and show that the $\mathfrak{T}$, and $\mathfrak{T}'$ are conjugate. Now $\mathfrak{L}^{(1)} = \mathfrak{T}\mathfrak{A} = \mathfrak{T}'\mathfrak{A}$ is either $\mathfrak{A}$ or $\{0\}$, by the minimality of $\mathfrak{A}$ and the condition $\mathfrak{A}^p = \{0\}$. If $\mathfrak{L}^{(1)} = \{0\}$, then $\mathfrak{T} = \mathfrak{T}'$ and there is nothing to prove. Thus, we may assume that $\mathfrak{A} = \mathfrak{T}\mathfrak{A} = \mathfrak{T}'\mathfrak{A}$. For $s \in \mathfrak{T}$, let $s = m + f(s)$ with $m \in \mathfrak{T}'$ and $f(s) \in \mathfrak{A}$. For $s, t \in \mathfrak{T}$, $0 = st$ leads to $f(t)s = f(s)t$. Choose $x \in \mathfrak{T}$ such that $\mathfrak{A}_0(\operatorname{ad} x) = \mathfrak{A}_0(\operatorname{ad} \mathfrak{T}) = \{0\}$, by 4.3.2, and choose $v \in \mathfrak{A}$ such that $vx = f(x)$. For $t \in \mathfrak{T}$, $(vt)x = (vx)t - v(xt) = (vx)t = f(x)t = f(t)x$. But $(vt)x = f(t)x$ implies that $vt = f(t)$, by the choice of x. Thus, $vt = f(t)$ for $t \in \mathfrak{T}$. We claim that $\mathfrak{T}e^v = \mathfrak{T}'$. Note first that for $s \in \mathfrak{T}$, $sE^{v} = s + sv = s - f(s) \in \mathfrak{T}'$ so that $\mathfrak{T}E^{v} \subset \mathfrak{T}'$. It follows that $\mathfrak{T}e^{v} \subset \mathfrak{T}'$, hence $\mathfrak{T}e^{v} = \mathfrak{T}'$ by the choice of $\mathfrak{T}$ as torus of maximal dimension.

4.6.9 Corollary
If $\mathfrak{L}$ is solvable and one Cartan subalgebra of $\mathfrak{L}$ is a torus, then every Cartan subalgebra of $\mathfrak{L}$ is a torus.

PROOF. Use 4.6.4 and 4.6.8.

4.6.10 Corollary
Let $\mathfrak{L}$ be solvable and $(\operatorname{ad} x)^{p} = 0$ for $x \in \mathfrak{L}^{\infty}$. Then any two Cartan subalgebras $\mathfrak{H}$, $\mathfrak{H}'$ of $\mathfrak{L}$ are of the same dimension, and there exist $x_i \in \mathfrak{L}^{\infty}$ such that $\mathfrak{H}' = \mathfrak{H}E^{x_1} \ldots E^{x_n}$.

PROOF. We first note that for $\mathfrak{H}$ a Cartan subalgebra of $\mathfrak{L}$ and $x \in \mathfrak{L}_a(\mathfrak{T})$ $(a \neq 0)$, $\mathfrak{T}$ being a maximal torus of $\mathfrak{L}$, $\mathfrak{H}e^{x} = \mathfrak{H}E^{x}$. For, under the present hypothesis, $(\operatorname{ad} x)^{p} = 0$, since $x \in \mathfrak{L}^{\infty}$, and one can easily verify that this implies that $\mathfrak{H}E^{x} = \mathfrak{C}_{\mathfrak{L}}(\mathfrak{T})E^{x} = \mathfrak{C}_{\mathfrak{L}}(\mathfrak{T}E^{x}) = \mathfrak{C}_{\mathfrak{L}}(\mathfrak{T}e^{x}) = \mathfrak{H}e^{x}$. Now we invoke 4.6.8, which implies that there exist $x_i \in \mathfrak{L}^{\infty}$ such that $\mathfrak{H}' = \mathfrak{H}E^{x_1} \ldots E^{x_n}$. Since $(\operatorname{ad} x_i)^{p} = 0$ for all i, the E^{x_i} are nonsingular and $\dim \mathfrak{H} = \dim \mathfrak{H}'$.

Appendix The Zariski Topology

The purpose of this appendix is to develop some elementary facts about the Zariski topology which are needed in parts of this book. Throughout, k is a field, V and V' are finite-dimensional vector spaces over k, and $\{e_i \mid 1 \leq i \leq m\}$ and $\{e_j' \mid 1 \leq j \leq n\}$ are bases over k for V and V', respectively.

A.1 Definition

A *polynomial mapping* from k^m into k^n is a mapping $f\colon k^m \to k^n$ given by $f(\alpha_1, \ldots, \alpha_m) = (f_1(\alpha_1, \ldots, \alpha_m), \ldots, f_n(\alpha_1, \ldots, \alpha_m))$ where $f_j \in k[X_1, \ldots, X_m]$ for $1 \leq j \leq n$.

A.2 Definition

A *polynomial mapping* from V into V' *with respect to* bases $\{e_i|1 \leq i \leq m\}$, $\{e_j'|1 \leq j \leq n\}$ for V, V' is a mapping $f\colon V \to V'$ that is a polynomial mapping with respect to some bases $\{e_i \mid 1 \leq i \leq m\}$, $\{e_j' \mid 1 \leq j \leq n\}$ for V, V' over k.

A.3 Definition

$k[V]$ is the set of polynomial functions from V into k.

A.4 Definition

For $f \in k[V]$, $V_f = \{v \in V | f(v) \neq 0\}$.

A.5 Definition

The *Zariski topology* for V is the topology for V having $\{V_f | f \in k[V]\}$ as the base of open sets.

The points of V with the Zariski topology are closed. Polynomial mappings from V to V' are continuous in the Zariski topology.

A.6 Proposition

Let k be infinite. Then any nonempty open subset of V is dense in V.

PROOF. It suffices to show that if U_1, U_2 are nonempty open subsets of V, then $U_1 \cap U_2$ is nonempty. Thus, choose nonconstant polynomials f_1, f_2 such that $U_i \supset V_{f_i}$ for $i = 1, 2$. Then $U_1 \cap U_2$ contains $V_{f_1 f_2}$. Since $f_1 f_2$ is nonconstant and k infinite, $V_{f_1 f_2}$ is nonempty. Thus, $U_1 \cap U_2$ is nonempty.

References

1 **Albert, A. A., and Frank, M. S.**
Simple Lie algebras of characteristic *p*. *Rend. Torino*, **14,** 117–139 (1954–1955).

2 **Allen, H. P.**
Jordan algebras and Lie algebras of type D_4. *J. Alg.*, No. 2, 250–256 (1967).

3 **Barnes, D. W.**
On Cartan subalgebras of Lie algebras. *Math Z.*, **101,** 350–355 (1967).

4 **Block, R. E.**
Determination of the differentiably simple rings with a minimal ideal. *Ann. of Math.*, **90,** 433–459 (1969).

5 **Borel, A., and Mostow, G. D.**
On semi-simple automorphisms of Lie algebras. *Ann. of Math.*, **61,** 389–405 (1955).

6 **Bourbaki, N.**
Groupes et Algèbres de Lie. Chap. I, Algèbres de Lie. Hermann, Paris, 1960.

7 **Bourbaki, N.**
Groupes et Algèbres de Lie. Chaps. IV-VI: Groupes de Coxeter et Systèmes de Tits; Groupes Engendrés par Réflexions; Systèmes de Racines. Hermann, Paris, 1968.

8 **Chevalley, C., and Schafer, R. D.**
The exceptional Lie algebras F_4 and E_6. *Proc. Natl. Acad. Sci.*, **36,** 137–141 (1950).

9 **Curtis, C. W., and Reiner, I.**
Representation Theory of Finite Groups and Associative Algebras. Interscience, New York, 1962.

10 **Dieudonné, J.**
On semi-simple Lie algebras. *Proc. Am. Math. Soc.*, **4,** 931–932 (1953).

11 **Humphreys, J. E.**
Algebraic groups and modular Lie algebras over fields of prime characteristic. *Mem. Am. Math. Soc.*, No. 71 (1967).

12 **Jacobson, N.**
A note on automorphisms and derivations of Lie algebras. *Proc. Am. Math. Soc.*, **6,** 281–283 (1955).

13 **Jacobson, N.**
Lie Algebras. Interscience, New York, 1962.

14 **Jacobson, N.**
Field Theory (Vol. III of Lectures in Abstract Algebra). Van Nostrand, New York, 1964.

15 **Jacobson, N.**
Triality and Lie Algebras of Type D_4. *Rend. Palermo*, **13,** 1–25 (1964).

16 **Jacobson, N.**
Exceptional Lie Algebras. Marcel Dekker, New York, 1971.

17 **Jenner, W. E.**
On non-associative algebras associated with bilinear forms. *Pac. J. Math.*, **10,** 573–575 (1960).

18 **Kreknin, V. A.**
Solvability of Lie algebras with a regular automorphism of finite period. *Dokl. Akad. Nauk SSSR*, **150,** 467–469 (1963).

19 **Mills, W. H., and Seligman, G. B.**
Lie algebras of classical type. *J. Math. Mech.*, **6,** 519–548 (1957).

20 **Mostow, G. D.**
Fully reducible subgroups of algebraic groups. *Am. J. Math.*, **78,** 200–221 (1965).

21 **Schafer, R. D.**
An Introduction to Non-Associative Algebras. Academic Press, New York, 1966.

22 **Seligman, G. B.**
On Lie algebras of prime characteristic. *Mem. Am. Math. Soc.*, No. 19 (1956).

23 **Seligman, G. B.**
On automorphisms of Lie algebras of classical type. *Trans. Am. Math. Soc.*, **92,** 430–448 (1959).

24 **Seligman, G. B.**
On automorphisms of Lie algebras of classical type II. *Trans. Am. Math. Soc.*, **94,** 452–482 (1960).

25 **Seligman, G. B.**
On automorphisms of Lie algebras of classical type III. *Trans. Am. Math. Soc.*, **97,** 286–316 (1960).

26 Seligman, G. B.
Modular Lie Algebras. Springer, New York, 1967.

27 Serre, J. P.
Lie Algebras and Lie Groups. Benjamin, New York, 1965.

28 Serre, J. P.
Algèbres de Lie Semi-simples Complexes. Benjamin, New York, 1966.

29 Smith, D. A.
On fixed points of automorphisms of classical Lie algebras. *Pac. J. Math.*, **14,** 1079–1089 (1964).

30 Steinberg, R.
Automorphisms of classical Lie algebras. *Pac. J. Math.*, **11,** 1119–1129 (1961).

31 Taft, E. J.
Orthogonal conjugacies in associative and Lie algebras. *Trans. Am. Math. Soc.*, **113,** 18–29 (1964).

32 Winter, D. J.
On groups of automorphisms of Lie algebras. *J. of Alg.*, **8,** 131–142 (1968).

33 Winter, D. J.
Solvable and nilpotent subalgebras of Lie algebras. *Bull. Am. Math. Soc.*, **74,** No. 4, 754–758 (1969).

34 Winter, D. J.
On the toral structure of Lie p-algebras. *Acta Math.*, **123,** 70–81 (1969).

35 Winter, D. J.
Cartan subalgebras of a Lie algebra and its ideals. *Pac. J. Math.*, **33,** No. 2, 537–541 (1970).

Index